Springer Tracts in Advanced Robotics 84

Editors

For further volumes:
http://www.springer.com/series/5208

STAR (Springer Tracts in Advanced Robotics) has been promoted under the auspices of EURON (European Robotics Research Network)

Mario Prats, Angel P. del Pobil, and Pedro J. Sanz

Robot Physical Interaction through the Combination of Vision, Tactile and Force Feedback

Applications to Assistive Robotics

Springer

Authors

Dr. Mario Prats
Dept. ICC
Jaume-I University
12006 Castellon
Spain
E-mail: mprats@uji.es

Prof. Pedro J. Sanz
Dept. ICC
Jaume-I University
12006 Castellon
Spain
E-mail: sanzp@uji.es

Prof. Angel P. del Pobil
Dept. ICC
Jaume-I University
12006 Castellon
Spain
E-mail: pobil@uji.es

ISSN 1610-7438 e-ISSN 1610-742X
ISBN 978-3-642-33240-1 e-ISBN 978-3-642-33241-8
DOI 10.1007/978-3-642-33241-8
Springer Heidelberg New York Dordrecht London

Library of Congress Control Number: 2012946918

Printed on acid-free paper

Springer is part of Springer Science+Business Media (www.springer.com)

To Arantxa, for all these years. Mario.

To Azucena, Iván and Sofía for their unconditional love and understanding. Angel.

To Victoria, for her invaluable support. Pedro.

Foreword

Robotics is undergoing a major transformation in scope and dimension. From a largely dominant industrial focus, robotics is rapidly expanding into human environments and vigorously engaged in its new challenges. Interacting with, assisting, serving, and exploring with humans, the emerging robots will increasingly touch people and their lives.

Beyond its impact on physical robots, the body of knowledge robotics has produced is revealing a much wider range of applications reaching across diverse research areas and scientific disciplines, such as: biomechanics, haptics, neurosciences, virtual simulation, animation, surgery, and sensor networks among others. In return, the challenges of the new emerging areas are proving an abundant source of stimulation and insights for the field of robotics. It is indeed at the intersection of disciplines that the most striking advances happen.

The *Springer Tracts in Advanced Robotics (STAR)* is devoted to bringing to the research community the latest advances in the robotics field on the basis of their significance and quality. Through a wide and timely dissemination of critical research developments in robotics, our objective with this series is to promote more exchanges and collaborations among the researchers in the community and contribute to further advancements in this rapidly growing field.

The monograph by Mario Prats, Angel del Pobil and Pedro Sanz and is a contribution in the area of controlling the physical interaction of a robot with the environment. A new framework for sensor-based physical interaction is introduced, allowing an integrated versatile planning of object grasping and bimanual manipulation tasks and their dependable execution. Force, vision and tactile feedback are incrementally introduced leading to a multimodal control which aims at mimicking human dexterity during interaction. Results are validated in a rich set of experiments on different mobile manipulator and humanoid platforms, revealing a promising outlook toward the goal of achieving autonomous manipulation.

Remarkably, the monograph is based on the first author's doctoral thesis, which received the prize of the Tenth Edition of the EURON Georges Giralt PhD Award devoted to the best PhD thesis in Robotics in Europe. A very fine addition to STAR!

Naples, Italy
July 2012

Bruno Siciliano
STAR Editor

Preface

Autonomous manipulation is one of the most important challenges in robotics. A robot with advanced manipulation capabilities would generate a great impact on the society, specially in what concerns service applications. Robots that autonomously manipulate objects commonly found in human environments have the potential to raise the quality of life for millions of people, including the elderly and physically disabled. There are three fundamental challenges for any mobile manipulator or humanoid robot intended to perform service tasks: *versatility*, *autonomy* and *dependability*. Versatility is defined as the capability to adapt to different situations, instead of being limited to a particular task. Autonomy concerns the level of independence in the robot operation: an autonomous robot should be able to perform its actions without human intervention. Finally, dependability involves different concepts such as reliability and safety, and refers to the capability of successfully completing an action even under important modeling errors or inaccurate sensor information.

A complete manipulation task normally involves two sequential actions: that of achieving a suitable grasp or contact configuration, and the subsequent motion required by the task. As reported by several anatomical studies during the 20th century, both aspects are highly related, in the sense that the task requirements play a very important role in the selection of the grasp, and, inversely, the task motion that can be performed after the contact depends completely on the forces that can be transmitted through them. This grasp-task interplay has received little attention in the robotics community. Grasp planning approaches usually consider general pick and place tasks. Similarly, task planning and constrained motion control research normally assumes that an appropriate contact configuration is available, and that it remains suitable during the overall execution of the task.

We claim that a unified treatment of grasp and task-related aspects would imply very important advances in the versatility, autonomy and dependability of robotic manipulation. Therefore, we advocate the introduction of task-related aspects into the classical knowledge-based grasp concept, leading to *task-oriented grasps*. In a similar manner, grasp-related issues are also considered during the execution of a task, leading to *grasp-oriented tasks*. Both task-oriented grasps and grasp-oriented tasks compose a unified representation which we call *physical interaction*. This new concept suppresses the classical boundaries between the grasp and the task, and introduces the new problems of physical interaction *specification*, *planning* and *execution*.

In this book, we develop the concept of physical interaction that has its foundations in the classical task frame formalism and the concept of grasp preshaping. We propose several contributions concerning those problems, namely the specification, planning and execution of physical interactions, unified under the name *FPI approach* (Framework for Physical Interaction). First, the *theoretical framework* for the integrated specification of physical interaction tasks is defined, supporting a great variety of actions, not only involving direct hand-object manipulation, but also the use of tools or bimanual manipulation. Next, the problem of *autonomous planning* of physical interaction tasks is addressed. From a high-level task description, the planner selects an appropriate task-oriented hand posture and builds the specification of the interaction task by using the previous framework. We then focus on the *dependable execution* of these tasks, and adopt a *sensor-based approach* composed of a grasp and task controller running simultaneously, and taking into consideration three different types of sensor feedback which provide rich information during human manipulation: force, vision and tactile feedback. Human dexterity relies on the simultaneous processing of these three perceptual modalities. The fact that the manipulation skills of existing robotic systems are still very far from those of humans can be partly accounted for by a lack of integrated multimodal approaches to the problem. We present our research on the combination of vision, tactile and force sensing for manipulation with robotic hands, as part of our FPI approach. Our results show that our multimodal controller outperforms the bimodal or single-sensor approaches.

We introduce numerous new methods and concepts, such as ideal task-oriented hand preshapes or hand adaptors, as part of our unified FPI approach to manipulation. The FPI approach provides important advances with respect to the versatility, autonomy and dependability of state-of-the-art robotic manipulation. The consideration of task-related aspects into the grasp selection allows to address a wide range of tasks far beyond those of pick and place. In addition, these tasks can be autonomously planned by the physical interaction planner, instead of adopting preprogrammed ad-hoc solutions. Finally, advances in dependability are provided by novel grasp-task sensor-based control methods using vision, tactile and force feedback, and exploiting joint and grasp redundancy. All these contributions are validated in the real world with several experiments using different robots placed on household environments, such as the UJI Mobile Manipulator at Jaume I University, the Armar III humanoid robot at the University of Karlsruhe, and the ISRC Mobile Manipulator at Sungkyunkwan University. These robots are capable of performing tasks such as door opening, drawer opening, or grasping a book from a full shelf. As just one example of this validation, work on the high-DoF humanoid robot Armar III demonstrates a control system that successfully operates unmodeled mechanisms with widely varying structure in a general way with natural motions.

The foundation of the research described in this book was Mario Prats' PhD Thesis. This work was recipient of various national and international awards, including the European 2010 Georges Giralt Award and the 2009 Robotdalen Scientific Award Honorary Mention. The book is also accompanied by a website `http://www.physical-interaction.org` that contains complementary materials.

Castellón, July 2012 Mario Prats, Angel P. del Pobil and Pedro J. Sanz

Acknowledgements

There is a large number of people to whom we would like to express our gratitude, starting with all the members of the UJI Robotic Intelligence Laboratory and department colleagues at Jaume I University (Castellón, Spain), where most of the research work described in this book has been conducted.

Furthermore, during the development of this work, we have collaborated with other research labs in France, South Korea and Germany. This joint work gave us the opportunity to meet different cultures and great people. In the first of Mario's stays abroad, Prof. Philippe Martinet and Youcef Mezouar welcomed him at LASMEA, in Clermont-Ferrand (France). We are very grateful to all the members of the lab, but especially to Philippe, who encouraged him to follow a PhD program in robotics.

We would like to express our gratitude to Prof. Sukhan Lee, with whom we worked at the Intelligent Systems Research Center, and the Interaction Science Department, both at Sungkyunkwan University in South Korea. Part of the experimental work presented in this book was done there. Thanks also to all the members of the lab for so many great moments.

Our third collaboration was with the IAIM team headed by Prof. Rüdiger Dillmann at Karlsruhe (Germany). We wish to thank him and Tamim Asfour for making it possible for Mario to stay there, and for granting access to the laboratory facilities. Thanks also to all the lab members and the administrative staff for their warm welcome.

We are very grateful to all our friends, and to our families, specially to our parents, wives and children, who gave us constant encouragement and tireless support.

Finally, this work would not have been possible without the financial support received by the authors and the UJI Robotic Intelligence Laboratory from different funding agencies such as Ministerio de Ciencia e Innovación (DPI-2008-06636 and DPI2011-27846), Generalitat Valenciana (PROMETEO/2009/052 and an FPI scholarship), Fundació Caixa Castelló-Bancaixa (P1-1B2008-51 and P1-1B2011-54), the WCU (World Class University) program of the National Research Foundation of Korea funded by the Ministry of Education, Science and Technology (R31-2008-000-10062-0), and the European Commission (grant agreement 217077, EYESHOTS project). The authors are also very grateful to different financial support received through the IRS-Lab, from different funding agencies such as Ministerio de Ciencia e Innovación (DPI2004-01920, DPI2008-06548-C03 and DPI2011-27977-C03), Generalitat Valenciana (ACOMP/2012/252), Fundació Caixa Castelló-Bancaixa (PI.1B2011-17), and the European Commission (grant agreement 248497, TRIDENT project).

The authors are also very grateful to the Lagadic team (Irisa/INRIA, Rennes, France) for their excellent ViSP library. We would also like to express our gratitude to Andrew Kator and Jennifer Legaz for their complete Blender model repository licensed under Creative Commons: http://www.katorlegaz.com/3d_models/. In general, thanks to the Open Source community because of so many professional software.

Contents

Chapter 1
Introduction

According to the World Robotics 2011 report [79], up to the end of 2014, more than 14 million service robots will populate the world. A *service robot* is a robot which operates partially of fully autonomously to provide services useful to the humans [10]. The group of service robots can be further divided into *professional service robots*, intended to perform service tasks at a professional level, and *personal robots*, suitable for education, assistance or entertainment tasks at home. Among these, it is of special interest the group of *assistive robots*, which main purpose is to help elderly and impaired people in their daily life.

It is well-known that the elderly population is considerably increasing in the industrialized countries, in percentage with the overall population. Figure 1.1 shows the eight countries with a higher estimation of elderly population in 2010 and 2050 [135]. It is expected that the percentage of people over 65 years old will almost duplicate in most of the industrialized countries by 2050. In the concrete case of Spain, estimations report that it will be the third country with more elderly people in 2050, only after Japan and Korea.

The elderly population constitute, together with impaired people, a group of the society that have to face with innumerable problems in their daily life. According to a report of the Spanish statistics institute [81], in the year 2008 a 23.19% of the

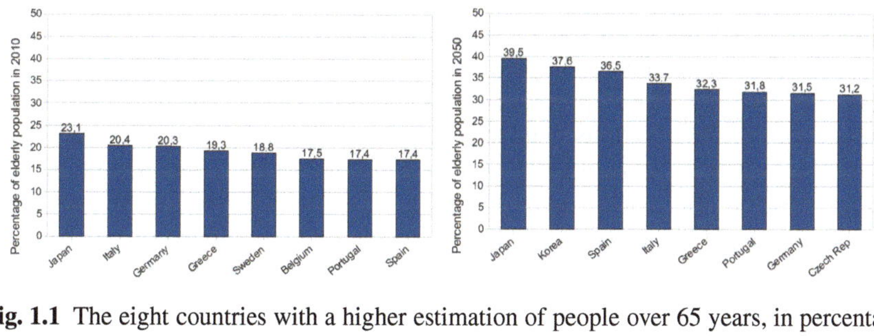

Fig. 1.1 The eight countries with a higher estimation of people over 65 years, in percentage to the overall population, in 2010 and 2050

M. Prats et al.: Robot Physical Interaction through the Combination of Vision, STAR 84, pp. 1–5.
springerlink.com

Spanish homes contained at least one disabled person (approximately 3.3 millions of the Spanish homes). In 608.000 of the cases the disabled person was living alone. The report also states that in approximately the 39.85% of the Spanish homes with disabled members, the impaired person is, in addition, more than 65 years old. The most common case of a home containing impaired people above 65 years is that composed of two persons, which represents a 40.48%, whereas in the 24% of the cases, the disabled and elderly person is living alone. The 74% of disabled people report difficulties to perform basic daily activities, such as personal hygiene, dressing, feeding, cooking, doing housework, moving around the home and performing simple tasks. Although this is the particular Spanish situation, these numbers seem to be in agreement with other industrialized countries.

Surprisingly, it is estimated that around the 26.6% of the impaired Spanish population (more than a million persons) does not receive any help for performing those tasks for which they have difficulties. This percentage highlights that current technical and personal assistance practices are insufficient. The high-costs of the health-care sector, together with the lack of independence that assisted people experiment, lead to living conditions which are still unsatisfactory for most of the elderly and impaired population. The society is now facing the challenge of how to effectively take care of these persons, providing low-cost assistance and increased comfort. Assistive robots may represent a valid solution.

Several surveys have reported the convenience of assistive robots for the people with necessities [174, 65, 136]. An assistive robot able to perform everyday tasks would allow the people with mobility and other problems to live independently and autonomously. Among the functionality that users would demand from an assistive robot, the most highly rated are: do housework, prepare drinks and food, reach books and other objects from a shelf, plugging things, loading a video, watering plants and other gardening tasks, get items from the refrigerator, turn knobs, open/close doors and drawers, turning appliances on and off, and operating light switches.

Current robotics technology is still very distant from reaching all of these functionalities [87], which in most (or all) of the cases require advanced manipulation capabilities, offering advanced levels of versatility, autonomy and dependability. There are still no autonomous assistive robots with manipulation capabilities (from now, *assistive manipulators*) that have reached enough maturity for becoming commercially available. Only the Manus arm [42] is being commercialized as a manipulator that can be attached to a wheelchair and controlled by the user with a joystick or other input device. However, some start-up companies have ambitious plans to improve and commercialize their mobile assistive robotic prototypes, as, for example, the *Personal Robot* project [191] of Willow Garage, or ReadyBot [153]. In other cases, collaborations between big companies and universities aim at the same goal, as it is the case of the *Home Assistant Robot* [64], which is being built by researchers at Tokyo University in collaboration with Toyota. If we move into research groups, most of the research on assistive robot manipulation has been focused on wheel-chair manipulators [13, 3, 42], although interest on multipurpose mobile manipulation and humanoid robots is increasing considerably. One example is the Care-O-bot II assistance robot, which was introduced in [59], and included a manipulator that allowed to perform fetch and carry task, as an extension to Care-O-bot I [166] which did not include any manipulation capabilities. Reference [132] described a mobile manipulator for performing assistive tasks inspired in service dogs, which was able to open doors and drawers by grasping colored towels tied to the handles. The STAIR (STanford AI Robot) project [151] is aimed at building a mobile manipulator that can navigate through home and office environments, assisting users in their everyday actions. Finally, four outstanding examples of assistive humanoid robots are the Armar-III robot [6], HRP-2 [85], Twendy-one [185] and Justin [17].

The lack of commercial assistive manipulators is in contrast with the number of other service robots which have already succeeded in other areas, such as robot nurses [125], museum guidance robots [172], robots at hospitals [51], blind-guiding robots [94], tele-presence robots [156], or entertainment robots [155, 77]. Some of these machines are already being commercialized (or have plans to) by companies like Neobotix, WowWee, Anybots, I-Robot, etc. Surprisingly, none of the commercialized service robots include manipulation capabilities, and most of them even come without arms and hands. One possible reason is that robotic manipulation has still not reached the required levels of versatility, autonomy and dependability that other areas of robotics have achieved. Whereas current mobile robots are able to navigate around the environment in safety conditions, robotic manipulation seems to be still in its early stages, and a long way from the capabilities found in primates. Versatile, autonomous and dependable robotic manipulation techniques are required that open the doors to the development of useful assistive robots.

It is thought that, in a long perspective, the assistive robotics market will be a key area in robotics, and even the emergence of a new industry [56, 60]. An assistive robot with manipulation capabilities would not only have direct positive implications for the care of the elderly and impaired people, but also on other areas such as remote surveillance, tele-presence, space exploration, etc. Current commercial surveillance robots used by security companies do not include manipulation capabilities. This fact highly limits the mobility of the robot, which cannot open doors for accessing to new spaces. The inclusion of advanced manipulation functionalities into these robots would allow remote operators to access new spaces through doors and elevators, or to switch off the lights, for example. Similarly, an assistive robot at home could be used remotely by the owner in order to operate home appliances, close/open windows, check the availability of an item, etc., or by a user with mobility limitations in order to bring a drink from the fridge, grasp a book, and so on.

The purpose of this work is to advance towards more versatile, autonomous and dependable manipulation. More concretely, we support that a further integration of two aspects that have been normally considered independently, the grasp and the task, could imply a breakthrough in robotic manipulation. Classical studies of the human prehension capabilities report that the intended task represents a very important aspect for the selection of a suitable grasp or contact configuration in humans. Conversely, the contact configuration on a given object introduces limitations in the number of tasks that can be performed with it, and it is of utmost importance that

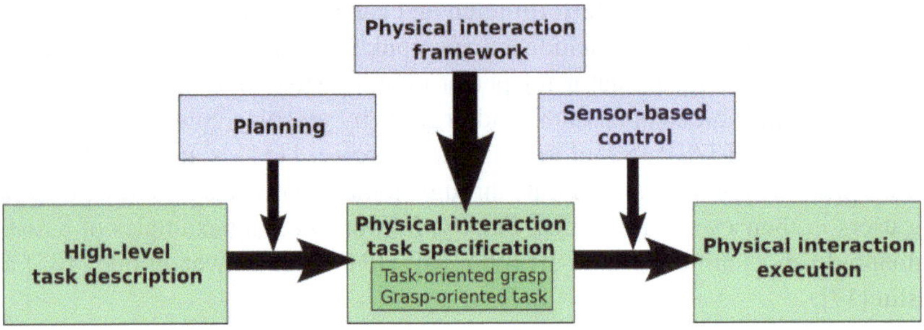

Fig. 1.2 A summary of the contributions of this book

the contact state remains suitable during the overall execution of the task. This interplay has been rarely considered in the robotics community. Research on grasp planning normally focuses on the particular task of pick and place. However, the kind of tasks that humans perform in their daily life goes far beyond pick and place tasks. Similarly, research on task planning and constrained motion control usually assumes that an appropriate contact configuration has been previously achieved and that it remains suitable during the execution. There is a lack of collaboration between grasp-based and task-based approaches.

In our opinion, this lack of grasp-task integration introduces important limits in the achievement of the desired properties of versatility, autonomy and dependability in robotic manipulation. Firstly, as the task requirements are not considered in the selection of the grasp, robot grasping capabilities are typically limited to general pick and place actions. Secondly, the lack of grasp planning inside task execution approaches normally leads to ad-hoc solutions which are only valid for one particular system and task. And, finally, the absence of grasp control processes during the task execution prevents the detection of changes in the contact configuration and its corresponding correction.

Therefore, we advocate for the introduction of task-related aspects into the classical knowledge-based grasp concept, leading to *task-oriented grasps*. In a similar manner, grasp-related issues are also considered during the execution of a task, leading to *grasp-oriented tasks*. Both task-oriented grasps and grasp-oriented tasks compose a unified representation which we call *physical interaction*. This new concept suppresses the classical boundaries between the grasp and the task, and introduces the new problems of physical interaction *specification*, *planning* and *execution*.

In this monograph, we propose several contributions in these lines, as depicted in Figure 1.2. First, a *theoretical framework* for the integrated specification of physical interaction tasks is defined. The framework is built on top of two well-established approaches to grasp planning and task specification: the knowledge-based approach, and the Task Frame Formalism. The link between the grasp and the task is established by means of the relationships between several coordinate systems called the *physical interaction frames*. The proposed framework allows for an integrated specification of a grasp and its subsequent task, and supports a great variety of actions, not only involving direct hand-object manipulation, but also the use of tools or manipulation with two hands. Next, the problem of *autonomous planning* of physical interaction tasks is addressed. From a high-level task description in terms of *object actions*, the planner selects an appropriate *task-oriented hand posture* and builds the physical interaction task specification using the previous framework. We then focus on the *dependable execution* of these physical interaction tasks, which is probably the most important part of this work. A *sensor-based approach* is adopted, and three different types of sensor feedback which provide rich information during human manipulation are considered: force, vision and tactile feedback. Several strategies for processing and combining the information coming from these sensors in different manipulation situations are proposed, including reliable force-based methods, a novel *vision-force control* approach, and a *hybrid vision-tactile-force controller*. Finally, all the contributions of this book are validated with experiments on real robots, including two different mobile manipulators and a humanoid robot. This work also presents the first steps towards a useful assistive manipulator: the *UJI Service Robot*. The number of different robots where these methods have been implemented, and the variety of tasks that have been considered, show the versatility of our approach, and its suitability for the autonomous and dependable execution of tasks under modelling errors and geometric uncertainties.

This book is organized as follows. Chapter 2 describes the state of the art in grasp and task control and identifies a lack of general approaches that consider the *grasp* and the *task* as related problems. The concept of *physical interaction* is then introduced as a more general approach including both aspects, and the methodology for developing such concept is explained. Chapter 3 establishes a conceptual framework that allows to specify physical interaction tasks with sensor-based control purposes, in terms of the *physical interaction frames*. Chapter 4 describes a *physical interaction planner* based on the previous framework and the new concepts of *task-oriented hand preshapes*, *object actions* and *object tasks*. Chapters 5, 6 and 7 explore different sensor processing methods and control strategies for the robust execution of physical interaction tasks, including, respectively, force-only, vision-force and vision-tactile-force control, paying special attention to its applicability in real situations. Chapter 8 focuses on the application of the framework, planner and control methods into a service robot prototype, and describes future challenges to be addressed as natural extensions to this work. Finally, the main contributions of this book, together with the conclusions and future work, are remarked in chapter 9.

Chapter 2
Robot-Environment Interaction

2.1 Brief Introduction

Generally speaking, a robot can physically interact with the environment by the sequential execution of two actions: that of achieving a suitable contact configuration, and the subsequent motion required by the task. Although contacts can be potentially performed with any part of the body, in this work we focus only on those tasks performed using the hand exclusively. The set of contacts generated by the hand on a particular object compose what we call a *grasp*. Therefore, throughout this book the *grasp* term adopts the general meaning of any hand-object contact configuration, either prehensile or non-prehensile, in contrast to the usual understanding of a grasp as a prehensile pattern. The term *task* is used to refer to the force that must be applied to and/or the motion that must be performed on an object after a suitable contact configuration has been established, in order to accomplish a particular goal. Grasp and task are two inseparable concepts: one precedes the other. The purpose of the grasp is to achieve a contact configuration which is appropriate for the task requirements.

This chapter reviews the state of the art in robot-environment interaction, including grasping and task-constrained motion control. We pay special attention to the interplay between the grasp and the task, and identify a lack of general methodologies for integrated grasp-task planning and control. We then introduce the new concept of physical interaction, as an integrated approach to manipulation taking into consideration grasp and task-related aspects. An outline of our approach for developing this new concept is presented at the end of this chapter, and detailed throughout the rest of the book.

2.2 Previous Approaches

Robotic manipulation of the environment is a topic that has been considered since the beginning of robotic research. Indeed, the first industrial robotic manipulators were implanted in 1961 at General Motors. From then, industrial robots have

M. Prats et al.: Robot Physical Interaction through the Combination of Vision, STAR 84, pp. 7–17.
springerlink.com

emerged as a strong market that has installed near one million units all over the world [78]. A robot can physically act on the environment either by grasping actions or by dynamic interaction with fixed and mobile objects, which constitute the two main important aspects of a robot-environment interaction: the grasp and the task. In this section, a bibliography review of both fields is presented, with the purpose of adopting well-established techniques that can be used for building a unified grasp-task representation.

2.2.1 Robotic Grasping

In our daily life, hands are used for restraining objects [11], also called *fixturing* or prehensile manipulation, for manipulating objects with fingers, known as dexterous manipulation [138], and for non-prehensile manipulation [128]. According to reference [129], prehensile manipulation is the action of holding an object, either fixed or free, by a gripping or pinching action. Non-prehensile manipulation, however, includes actions like pushing, lifting, tapping, etc. which do not necessarily require a firm grasp. Dexterous manipulation has been defined as an object-centered approach [138], which involves the cooperation of the fingers in order to apply a desired force-torque wrench on the object. Humans continuously combine the three different manipulation patterns in their daily life.

The three cases have been addressed in robotics, although more attention has been put to prehensile manipulation [11]. Two main approaches to object fixturing exist: the *contact-level approach* and the *knowledge-based approach.*

2.2.1.1 The Contact-Level Approach to Grasping

The contact-level approach considers the grasp as a set of contacts, each one transmitting a force-torque to the object. The purpose of contact-level grasp synthesis algorithms is to find a set of contacts so that the space of all the possible forces-torques that can be applied to the object through the contacts, contains the space of all the possible disturbance wrenches.

Several contact models have been developed, being Salisbury PhD thesis [159] the reference source. Salisbury proposed eight different kinematic models of contact. Among all the contact models included in his taxonomy, three of them have been widely considered: *frictionless point contact* (FPC), *point contact with friction* (PCWF) and *soft finger contact* (SFC). A frictionless point contact can only transmit a force along the normal to the object surface in the contact point. Point contact with friction can also transmit small forces in the tangential directions to the surface at the contact point. Soft-finger contact considers an additional torque around the contact normal, being the most realistic case.

In contact models considering friction, the *friction cone* is introduced as a geometric representation of the frictional force limits. In order to compute the total wrench applied to the object through the contacts, the individual contact wrenches must be combined. Due to the non-linearity of the friction cone, this problem requires intensive computational effort, and several approaches have been proposed for efficiently dealing with it. The authors in [63] applied convex optimization techniques to a set of linear matrix inequalities built by linearizing the contact cones. Reference [193] proposed an algorithm for computing grasps of any number of contacts on 3D objects, based on the Q-distance measure. In most of the cases, the frictional cone is linearized by approximating it with a polyhedral cone defined in vector space [15]. This allows to define the *Grasp Wrench Space* concept (GWS), as the vector space that is spanned by all the possible contact forces lying inside the friction cone at each contact.

Similarly, the *Task Wrench Space* concept (TWS) was introduced for modeling all the possible wrenches that are expected to occur on the body during the task [102, 53]. Reference [148] introduced the *Object Wrench Space* (OWS) as a more natural way to describe the TWS. The grasp planning problem is then formally addressed as finding a set of contacts which generate a GWS containing the TWS, and which maximizes a quality measure.

Most approaches consider the TWS to be a sphere (or a circle in the 2D case), in order to be able to transmit any force to the object, and compensate any external disturbance. This property is known as *force-closure*, and has become the standard for evaluating the grasp quality [11]. The authors in [53] proposed a force-closure quality measure interpreted geometrically as the radius of the largest sphere which is completely contained in the GWS. Using this measure, the authors defined a force-closure grasp planning algorithm and two optimization measures, one for minimizing the maximum finger force, and another for minimizing the total finger force, using a two-jaw parallel gripper on 2D polygonal objects. This measure has also been adopted in [14], that proposed the Incremental Convex Hull Calculation method for computing the GWS efficiently, and in [15] for efficient grasp planning, where random potential grasps were generated and evaluated for finding a good (although not optimal) grasp.

2.2.1.2 The Knowledge-Based Approach to Grasping

Some researchers have argued that pure contact-level approaches usually do not take the hand constraints into consideration [190, 119, 74], producing a set of contacts which are not reachable in practice. In addition, even if the hand is able to reach the contact points, it is difficult to perform such grasps in practice, mainly because of the lack of accurate sensors and errors in the models and robot-object positioning [121]. This argument is supported by the small number of articles that apply the contact-level approach in real situations.

An alternative approach is to grasp with predefined hand postures, which is known as the *knowledge-based approach* to grasping [175, 138], or also the *qualitative approach* [121]. The original idea is to consider a set of predefined prehensile patterns and classify them according to its suitability for the object geometry and for the task. The grasp planning problem is then addressed as the selection of the grasp preshape which best adapts to the particular geometry and task.

The concept of hand preshapes was firstly introduced in several studies of the human prehension capabilities in [164], [179] and [128], and later applied to robotics in [105] and [31], that defined a taxonomy of human grasps with applications to manufacturing of robotic hands. Since then, several papers have adopted this approach for grasping. Reference [175] applied hand preshapes to the Salisbury hand. The concept of *grasp strategy*, consisting of a hand preshape extended with a set of digit trajectories was introduced in [190]. The authors in [120] proposed to approximate the object shape with a set of shape primitives, and provided a planning algorithm for the Barrett Hand, by defining the most suitable preshape for each primitive. A combination of model-based and appearance-based methods for visual recognition and localization of objects was used in [121]. Object models were later used as input for a grasp planner that selected the most suitable preshape for a humanoid hand. References [75] and [74] developed vision methods for approximating an unknown 3D shape with several box primitives, over which a grasp was planned. In contrast to contact-level methods, the knowledge-based approach to grasping has shown to be an intuitive and practical solution which has lead to numerous applications in real situations.

2.2.1.3 Task-Oriented Grasping

Some researchers have reported the importance of designing task-oriented grasping algorithms, defined as those which take into account the task requirements. Inspired by anatomical studies of the human prehension capabilities, Cutkosky reported that the choice of a grasp depends mainly on the task to be performed [33], even more than on the object geometry. According to [102], stability is only a necessary condition for a good grasp, but not sufficient: a good grasp should be task-oriented, i.e. able to generate body wrenches that are relevant to the task. Therefore, force-closure is a too restrictive condition when task-oriented grasps are considered, because there is no need to resist all the possible external wrenches, but only those required by the subsequent task.

The synthesis of task-oriented grasps presents three main advantages. First is that a task-oriented grasp normally needs less number of contacts than a force-closed grasp, and thus, less number of fingers. The reason is that there is no need to resist all the possible forces, but only a small subset composed of the expected ones. The second advantage is that a task-oriented grasp is more *effective* than a force-closed grasp, for a given particular task. The reason is that contacts are focused to generate forces for the task, and not for unnecessary directions. Finally, a grasp planner taking into consideration the task constraints increases the robot versatility, as it allows to consider a wide range of actions far beyond pick and place.

The process of manipulating a part without a force- or form-closure is also known as non-prehensile manipulation [115], or *graspless* manipulation [1]. The difference with respect to task-oriented grasps is that some task-oriented grasps can also be prehensile. Non-prehensile manipulation has received little attention in the literature, although its importance in daily-life manipulation is unquestionable. Several studies exist, mainly in dynamics of non-prehensile manipulation [115, 104], stability measures [107], and motion planning of pushing, sliding, tumbling and pivoting tasks [194, 108]

Unfortunately, the task-oriented properties of a grasp have only been considered as quality measures, and following the contact-level approach. The first task-oriented grasp quality measure was presented in [102]. The authors introduced the concept of *task ellipsoid* for modeling a non-spherical TWS, and defined a task-oriented grasp quality measure as the radius of the largest task ellipsoid which is contained in the GWS. However, efficient methods for computing this measure were not proposed until recently. Borst et al. [16] proposed a method to compute the task ellipsoid from the TWS, and to scale the GWS and TWS in order to transform the ellipsoid into a sphere, thus representing the problem as a "sphere fitting" problem, which can be solved efficiently. The work in [66] approximated the task ellipsoid with a polytope, employing a set of task wrenches along the principal axis of the ellipsoid. Even though these approaches provide significant advances in task-oriented grasping, none of them have been applied in real robotic systems, probably because its contact-level nature and the associated difficulty to implement these methods in real situations.

Although the original description of prehensile patterns in robotics [31, 33] remarked the importance of the task requirements, the knowledge-based approach to grasping has still not considered the task-oriented nature of the grasp in practical works. Normally only pick and place actions are considered, together with the object geometry, for the selection of an appropriate prehensile pattern [74, 121]. Some works have also considered the constraints imposed by the hand kinematics [122].

2.2.2 *Robotic Task Execution*

A robot task can be defined as a goal arrangement of physical objects [96], or, more generally, as accomplishing a relative motion and/or controlled dynamic interaction between objects [37]. The later definition includes actions which do not necessarily involve object motion, such as pushing on a fixed surface. In general, performing a manipulation task involves dealing with position and force constraints in the robot end-effector.

Research in robotic task execution can be divided into global and local approaches. The global approaches have been usually adopted by the motion planning communities [96], which address the task execution problem as planning a joint space path for a high dimensional kinematic chain performing a desired end-effector motion in an environment with obstacles. In contrast, the local approaches, normally adopted by the control communities, aim at computing an arm-hand control signal based on sensor information, for an instantaneous, possibly constrained, hand motion [114, 90]. Whereas the former are suitable for finding global and optimal solutions, they normally require a perfect knowledge of the environment and intensive computational effort. In contrast, the local approaches are suitable when limited knowledge is available. The general procedure is to compute instantaneous joint motion based on sensor information, although they are subject to local minima and suboptimal motion. The main contributions of both approaches are outlined below.

2.2.2.1 Global Approaches: Task Planning

Motion planning is the problem of finding a collision-free joint space path for a given operational space task [96, 147, 62]. In the field of robotic manipulation, the desired end-effector trajectory is considered as the operational space task, whereas the joint space path is composed by a position-velocity trajectory for each actuator. Thus, the problem of motion planning for robotic manipulators can be seen as finding the joint control references which generate a desired end-effector trajectory and avoid robot collision with the environment.

Probabilistic Roadmaps [187] and *Rapidly-exploring Random Trees* [98] have become the most adopted algorithms for motion planning. Both of them provide feasible solutions based on probabilistic methods, even for kinematic chains with high number of DOF's. However, they are not suitable for dealing with closed kinematic chains or end-effector constraints in its original description. In order to deal with closed kinematic chains, the *Randomized Gradient Descent* algorithm was introduced in [97], and later extended in [192] for its use with general end-effector constraints in a case-by-case basis. The work in [177] adopted the Task Frame Formalism [114, 19] for unifying the representation of constrained motion and proposed two efficient methods for joint space path planning subject to task space constraints. The authors in [39] also made use of the Task Frame, and formulated the problem in order to allow relaxation of the task constraints by means of caging grasps.

Although planning methods can provide global and optimal solutions, they are computationally expensive, and its implementation in real environments is difficult, mainly because of the amount of previous knowledge about the environment that they require.

2.2.2.2 Local Approaches: Task Control

At the control level, several concepts have been developed in order to assist the programmer in the task specification and control problem. Mason introduced the concept of *compliance frame*, as a coordinate system related to the task and aligned with the object *natural constraints* [114]. Later it was also called *task coordinate frame* [67], and defined as an imaginary entity defined for simplifying the specification of manipulation motion. *Position-force hybrid control* was introduced for the implementation of different control modes on each frame direction [152]. Khatib introduced the concept of *generalized task specification matrix* [90], to properly divide the cartesian space into position and force-controlled directions according to the task, and developed the *operational space formulation*, where the overall dynamic control of the robot is decomposed into *task behavior* and *posture behavior*. Bruyninckx et al. formalized the original Mason's concepts into the *Task Frame* (TF) and *Task Frame Formalism* (TFF) [19], which has been the most accepted methodology for specifying sensor-based control tasks. The TFF has been adopted, for example, in [7] and [84] for vision-force control tasks, or in [106] for assembly tasks. De Schutter et al. realized that the TFF only applies to some task geometries, for which control modes can be assigned independently to each direction, and developed a more general approach, based on *feature frames*, where control references and constraints can be assigned to arbitrary cartesian directions, and where several types of geometric uncertainty can be taken into account [37].

Several authors have encapsulated the TF as part of *skill primitives* [67], which are basic parametrized robot motions that can be combined forming *skill primitive nets* [182, 55]. Some control architectures have included this concept [106, 93], which has been specially applied to assembly tasks [126].

2.2.3 Critical Discussion

Robotic manipulation is still far from the robustness and versatility offered by human infants, or even primate capabilities. Even though decades of research in robotic grasping and arm control have lead to fruitful results, very few of the robots currently found in research labs are able to perform tasks that go beyond pick and place in a dependable an autonomous manner. In the last years a remarkable number of mobile manipulators and complex humanoid robots have been developed, some of them endowed with advanced arms and hands. However, even though most of these robots have been designed with the purpose of assisting people at their homes, effort seems to be focused in locomotion, leaving manipulation almost unaddressed.

Several initiatives for building a multipurpose assistive mobile manipulator exist. One example is the PR-1 robot prototype [191], developed by the Willow Garage company in its Personal Robotics Program. This robot is the first of a series of personal robots designed for assisting people in human environments, although only tele-operated solutions have been presented until now. The STAIR (STanford AI Robot) project [151] is aimed at building a mobile manipulator that can navigate through home and office environments, interacting with objects [163], and with the ability of opening doors [146]. The El-E helper robot, developed at Georgia

Institute of Technology, mimics the capabilities of service dogs by grabbing hold of a towel to manipulate doors and drawers [132]. The same group was previously inspired by helper monkeys for grasping [88]. Care-O-bot II, designed for elderly care, included a manipulator that allowed him to perform vision-based fetch and carry tasks [59]. Although research in humanoid robots is mostly focused in locomotion aspects, some of them have also been used for manipulation purposes. Two outstanding examples are the Armar-III robot [6] and the HRP-2 [85]. Most of the previous robots are part of a long-term project for building a useful robotic assistant. However, versatility, autonomy and dependability seem to be three important limiting factors.

Among the applications that have been addressed in the literature, it is worth mentioning solutions for cleaning [112], contour following [7], assembly [182], and, especially, door opening [134, 143, 154, 140, 91, 146]. However, most of the published applications only provide specialized controllers for specific actions, which cannot easily scale for dealing with different tasks or systems. Furthermore, in some cases the robot reliability is highly limited by the lack of enough sensor feedback.

In our opinion, one of the reasons of the small number of robots actually using their hands for performing physical interaction tasks in real environments is the lack of a well-established methodology allowing for both grasp and task specification, planning and real-time dependable control. The grasp-task connection has been rarely considered in the literature, although it is unquestionable that the task requirements play a very important role in the selection of the grasp, and vice versa, the grasp state is of utmost importance during the task. There are different approaches to grasp planning, task planning and sensor-based control, but they have never been considered as a related problem, nor addressed from a common framework. There is the need of a more general approach to manipulation, where the grasp and the task are jointly considered, in a global framework based on multi-sensor information for real-time and real-life dependable physical interaction, providing current manipulators with advanced capabilities similar to those already existing in locomotion.

2.3 Our Approach

2.3.1 Task-Oriented Grasps and Grasp-Oriented Tasks

We advocate for the introduction of task-related aspects into the classical knowledge-based grasp concept, leading to *task-oriented grasps*. In a similar manner, grasp-related issues are also considered during the execution of a task, leading to *grasp-oriented tasks*. The consideration of task-related aspects into the grasp selection allows to address a wide range of tasks far beyond those of pick and place, thus increasing the robot versatility. In addition, the supervision and control of the grasp configuration while performing the task motion, allows to detect grasp problems

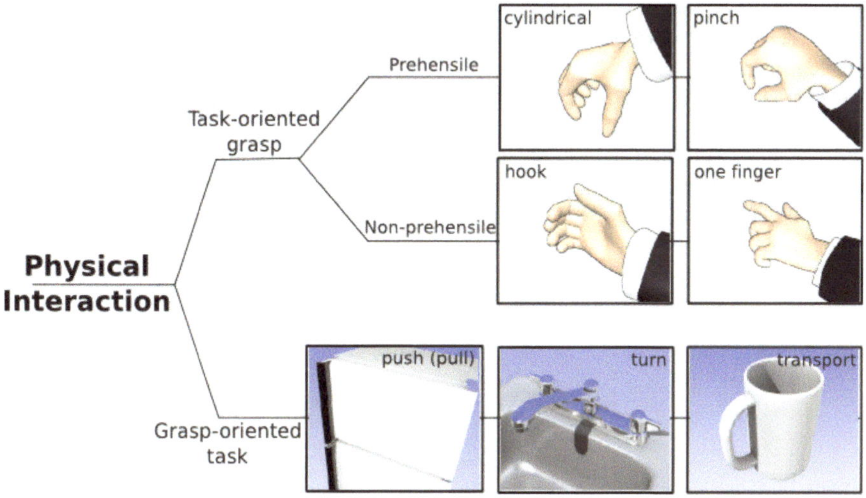

Fig. 2.1 The term *physical interaction* is introduced to refer indistinctly to the grasp (prehensile and non-prehensile) and to the task

and correct them accordingly, thus increasing the robot dependability. Finally, the integrated consideration of task-oriented grasps and grasp-oriented tasks covers the two aspects of a robot-environment interaction and allows to increase the robot autonomy.

2.3.2 The Concept of Physical Interaction

Both task-oriented grasps and grasp-oriented tasks compose a unified representation which we call *physical interaction*. Therefore, physical interaction is a term that will be used throughout this book to refer indistinctly to the grasp and the task aspects of a robot-environment interaction, as depicted in Figure 2.1. Generally speaking, it includes any kind of action involving a force transmission between the robot and the environment. Throughout this document, a *task-oriented grasp* (or simply *grasp*, for the shake of readability) is understood as any set of contacts between the hand and the object, either constraining the object motion in all the directions, known as *prehensile grips* in the taxonomy of [33], or constraining only some degrees of freedom (DOF's) in the case of *non-prehensile grasps*. The former group of grasps is the most widely considered in the literature because of its suitability for object lifting and transport actions. Non-prehensile grasps have received less attention, but are equally important for performing the most common tasks, such as pushing or turning actions. In this document, both the grasp and the subsequent task will be addressed without distinction as *physical interaction tasks*.

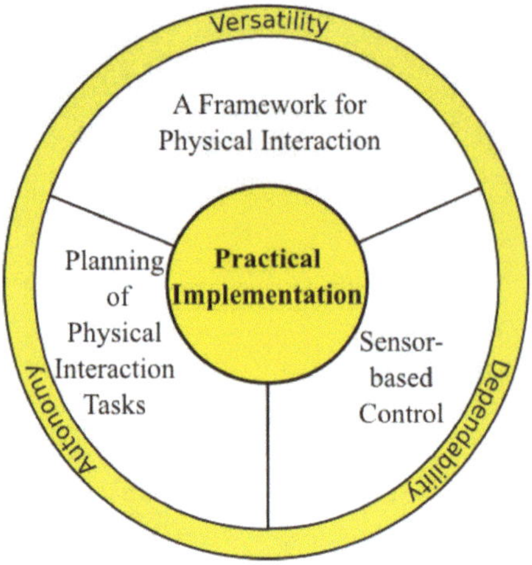

Fig. 2.2 This book concerns dependable practical implementation of physical interaction tasks, on the basis of a novel framework for grasp-task specification, task-oriented planning algorithms and robust control based on multi-sensor information

2.3.3 Methodology

In this book the physical interaction concept is developed, by addressing the following fundamental questions:

- How can everyday physical interaction be specified in a common framework, including both the grasp and the task, and supporting sensor-based control?
- How can a robot autonomously plan a physical interaction task, making use of this framework?
- What sensors are necessary, and how can a robot combine those sensors and control its motors for performing physical interaction tasks in a robust manner?

The questions above are addressed from a practical point of view, involving several real scenarios and fundamentally different robots and sensors. The ultimate goal is to develop a practical framework, appropriate for real-life experimentation, and to show its suitability for versatile, autonomous and dependable sensor-based execution of physical interaction tasks in human environments.

Figure 2.2 shows a chart composed of the main aspects addressed throughout this book:

- **A Framework for Physical Interaction**, which is our approach to the new problem of grasp-task specification for robust sensor-based execution. The proposed framework is based on well-established methods and allows for the specification of a large number of common physical interaction tasks, such as grasping parts, opening doors, pushing buttons, using tools, etc.
- **Planning of physical interaction tasks**, where algorithms for automatic grasp and task planning are developed in terms of the proposed framework. The tasks involve interaction with common objects and articulated furniture commonly found in household environments, such as doors, drawers, etc.
- **Sensor-based control**, which is the most extensive part of this book, and where new multi-sensor integration approaches in terms of the proposed framework are

proposed. Vision, force and tactile sensors are combined for performing up to three different types of robot motion: one, in charge of performing the grasp and keeping it stable during the task; another one, dedicated to perform the task motion; and a third one, in charge of adopting a comfortable arm posture according to a secondary task. More concretely, a novel vision-force combination scheme is developed and used in a control law for reaching an object and performing the task, while supervising the grasp at the same time. In addition, vision-force-tactile control is also addressed, in order to perform compliant physical interaction, even when important errors and uncertainties are present. Joint and grasp redundancy are also considered in order to perform the task with a comfortable arm posture, and increasing the robot workspace, respectively.
- The **implementation** of several common physical interaction tasks, in terms of the previous points, under three completely different robotic systems, including a humanoid robot and two different mobile manipulators. The purpose is to show how the proposed techniques can be successfully applied in real environments with essentially different hardware.

Throughout this work, the interplay between the different parts leads to several practical examples, which are validated with real robots, in real environments. As all the applications are built on top of a common framework, most of the code can be reused for many different tasks, allowing for fast implementation in very different hardware. In addition, both the grasp and the task are performed by combining several sensors, enabling dependable physical interaction, even under important errors and uncertainties inherent to real life experimentation. Our purpose is to provide the robotics community with new approaches that can be easily adopted in order to furnish our service robots with physical interaction capabilities similar to those already existing in terms of locomotion.

Chapter 3
A Framework for Sensor-Based Physical Interaction

3.1 Brief Introduction

The grasp and the task are two actions closely related. A grasp is always performed with a purpose in mind, and the execution of a task requires a suitable task-oriented grasp in most of the cases. The grasp does not need to be prehensile. Most of our daily life manipulation is done trough non-prehensile grasps. Interaction with the environment is always preceded by a contact configuration, as a result of a prehensile or non-prehensile grasp. And vice versa, a contact configuration on one object is always followed by a desired motion according to the task: lift, push, etc.

The most significant advances on the reciprocal relationship between the grasp and the task have been described in the previous section. Although the grasp-task interplay in our daily life is unquestionable, very few research has been performed in this line in robotics. Task planning and grasp planning communities have worked independently, and its intersection has received very little attention.

In order to fill this gap, we propose to include the grasp and the task into the more general concept of *physical interaction task*. This chapter presents a framework that allows the specification of physical interaction tasks, i.e., the integrated specification of the grasp and the task. For this, we adopt the most successful approaches to grasp and task specification, and extend them with additional elements that allow to establish a grasp-task link. As can be concluded from the previous section, these approaches are the Task Frame Formalism in the case of task specification, and the knowledge-based approach in the case of grasping.

First, the fundamental concepts of these methods are presented. The link between both approaches is studied, leading to a new method for grasp-task specification, oriented to sensor-based control. More concretely, a set of auxiliary frames (the *physical interaction frames*) is introduced. The physical interaction task is described in terms of the relationships between these frames. We provide a general control framework for the grasp and the task in terms of the physical interaction frames and

M. Prats et al.: Robot Physical Interaction through the Combination of Vision, STAR 84, pp. 19–38.
springerlink.com

study how sensors can be used for tracking the relationships between them. Finally, we show how the framework can also be applied to tasks involving the use of tools and two-hand actions.

3.2 The Task Frame Formalism

3.2.1 Concept

Compliant motion is a concept that refers to the motion of a robot manipulator when it is constrained by the task geometry. Opening a drawer, turning a door knob or polishing a surface are all examples of compliant motion tasks where the robot motion is constrained by a prismatic joint, a revolute joint and a planar contact respectively. In general, any compliant motion involves the lost of some degrees of freedom at the end-effector.

A robotic manipulator intended for compliant motion needs a task representation and specification approach that allows the programmer to define the task. The most accepted specification formalism for compliant motions tasks is the *Task Frame Formalism* (TFF).

The first contribution towards the TFF was proposed in [114], motivated by the increasing interest of finding an automatic method for the synthesis of control strategies for compliant motion, specially for automatic assembly applications. However, Mason only introduced the basic concepts of the formalism. It was in [36] and [19] where a a clear and formal description of the TFF was proposed, including practical examples and reporting its limitations. This formalism establishes an intuitive approach to model a motion constraint, and to specify the desired forces and motions in a compatible and controller-independent manner.

In the original description of the TFF given in [114], a cartesian coordinate system, later known as *compliance frame*, or *Task Frame* (TF), was introduced as a way to model the task geometry. The TF is composed of compliant and non-compliant axes, which indicate force-controlled and position- or velocity-controlled directions respectively. The position-controlled directions correspond to the task degrees of freedom. The task is represented as two rigid bodies, connected each other by a joint or a contact. Mason considered revolute, cylindrical, prismatic, spherical and screw joints, in addition to planar contacts. Each kind of joint imposes a different constraint on the relative velocity between the elements, and can be modeled with an equation system. These equations compose the *natural constraints*, and indicate force and velocity restrictions at the end-effector. For modeling the natural constraints, a geometric model of the task is needed. The TF has to be chosen so that, in the ideal case of no friction, force equations are decoupled from the velocity equations.

A more formal description of the TFF was given in [19], based on their extensive previous work on sensor-based programming and control [36, 35]. They defined three requirements and three options for modeling a task using the TFF concepts, and provided numerous examples of common applications. The main aspects of the TFF theory are outlined in the following section.

3.2.2 Details

Without considering the dynamic effects (i.e. from a kinetostatic point of view), constrained motion can be modeled as the relative instantaneous motion between two objects which does not generate any force other than frictional forces. One example of constrained motion could be the relative rotation between a door knob and the door itself, through the revolute joint that links them. Let $\mathbf{v} = (v_x, v_y, v_z, w_x, w_y, w_z)^T$ be a kinematic screw representing the relative velocity between the two linked objects, and $\mathbf{f} = (f_x, f_y, f_z, m_x, m_y, m_z)^T$ a wrench containing the forces and torques generated through the joint constraints, a constrained motion in a frictionless environment must satisfy the following expression, known as the *reciprocity condition*:

$$\mathbf{v}^T\mathbf{f} = 0 \tag{3.1}$$

All the possible kinematic screw vectors holding the reciprocity condition compose the *twist space*. Similarly, the *wrench space* is composed of all the possible wrench vectors reciprocal to all the velocities. For any kind of joint, it holds true that the twist and wrench spaces are disjoint sets. In addition, their union compose a six-dimensional vector space.

The TF is an orthogonal basis that allows to specify all the possible reciprocal vectors. It must be set by the task programmer according to the task geometry, so that part of its axes (the velocity-controlled directions) are a basis for the twist vector space, and the rest of its axes (the force-controlled directions) are a basis for the wrench vector space. According to [19], this is a requirement, called *Geometric Compatibility*, that any TF should satisfy, although it can be violated in some cases for the shake of simplicity in the specification.

An elementary task can be described, in terms of the TF, as a set of velocity references on the velocity-controlled axis, and force references on the force-controlled directions. The *Causal Compatibility* requirement ensures that the task specification is compatible with the constraint model specified by the TF, i.e., it must be possible to appropriately specify the task in terms of velocity and force vectors lying in the twist and wrench spaces spanned by the particular selection of the TF.

The TF must always remain geometrically compatible during the task execution, as considered by the *Time-invariance* requirement. This may require *tracking* the TF position and orientation during the task. Three different types of tracking can be considered:

No tracking: the particular geometry of the task allows to perform it completely while remaining geometrically compatible.

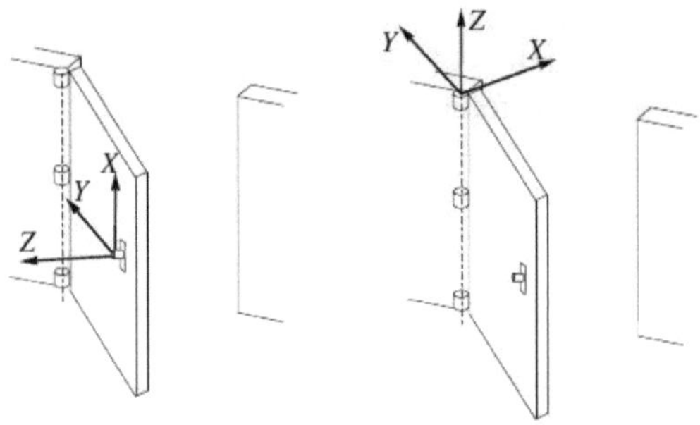

Fig. 3.1 Two valid examples of a Task Frame, under the Task Frame Formalism, adopted from [19]. For a door opening task, the task frame can be set either to the handle (left) or to the hinge axis (right). In the first case, both position and orientation of the TF change over time, whereas in the second case only orientation is modified. The first case violates the *Geometric Compatibility* requirement for the shake of simplicity in the specification.

Model-based tracking: a motion model is used to compute the TF displacement.
Sensor-based tracking: the new TF position and orientation is computed by using the robot sensors.

In this book we will focus on sensor-based tracking of the TF, with the purpose to avoid the use of motion models whenever possible. More concretely, we will study how vision, force and tactile sensors can be used for tracking a set of frames included in the grasp-task specification.

It is worth noting that the selection of the TF is not unique. The task programmer, or task planner, must chose the best placement according to the task requirements and the available geometric information. Bruyninckx et al. [19] introduced three criteria for selecting the best placement for the TF depending on the particular case. An example of two valid TF configurations for a door opening task can be found in Figure 3.1.

Finally, the TFF fits very well with the position-force hybrid control approach [152], which separates the space of controlled directions into position- or velocity-controlled and force-controlled subspaces. A straightforward mapping between the task specification and a hybrid control strategy is to impose velocity control to the TF velocity-controlled directions, and active force control to force-controlled directions. However, other possibilities exist, as the *Operational Space Formulation* [90], which is able to deal with manipulator dynamics and joint redundancy.

3.2.3 Limitations and Alternatives

Although a considerable number of manipulation tasks can be specified with the TFF, [20] reported that some particular task geometries are not considered by the original approach, specially those involving multiple contacts with non-orthogonal normals. In addition, the authors also reported that the TF, in its original description, is only a control-oriented concept, without any kind of symbolic information supporting intelligent processing.

As a more powerful alternative, they introduced the *Feature Frame Formalism* (FFF), built on *feature primitives*, which add symbolic information to assist the agreement between the task specification and the ongoing task execution. Under this approach, a general motion constraint is the combination of one or more *primitive motion constraints* defined on *feature primitives*. A feature can be a point contact, a line contact, a plane contact or an helical joint (which include the prismatic and revolute joints). As in the TFF, the choice of the particular feature frames affects to the specification complexity, although the physical meaning remains invariant. Bruyninckx et al. [20] suggested several guidelines for appropriately setting the feature frames depending on the joint type: for contact features, the origin can be set at the contact point, with one axis aligned with the contact normal; for a joint, the origin can be set somewhere in the joint axis, with a frame axis aligned with it. The TFF generalizes the task to be a point contact at the end-effector transmitting one force-torque to the environment, whereas the FFF can potentially deal with any number of contacts and, therefore, it allows to specify more complex tasks.

One of the most important aspects during constrained motion is geometric uncertainty. Errors in the models or robot positioning can generate considerable interaction forces. Since the TFF does not specify any method for the observation and identification of the geometric uncertainty, sensor-based control methods, specially those based on force feedback, are required for ensuring a safe execution. Alternatively, [37] extended the FFF for automatically dealing with the geometric uncertainties in the modeling of the feature frames. This method constitutes an important breakthrough, since it is possible to perform an integrated management of the constraints and the geometric uncertainty. However, a considerable task specification effort is necessary, thus making the implementation of automatic planners difficult.

Although the FFF provides a more powerful approach to task execution, our framework for describing physical interaction tasks is based on the TFF, mainly because of the following reasons:

- The TFF is a simple and intuitive approach to task specification, but still powerful.
- The TFF is suitable enough for the kind of tasks required from a home assistive manipulator. Most of these tasks involve interacting with home appliances and articulated furniture, which normally require acting on prismatic and revolute joints.
- The TFF is suitable for autonomous planning of physical interaction tasks. A high-level task description on an articulated object can be easily transformed automatically into a TFF-based specification.

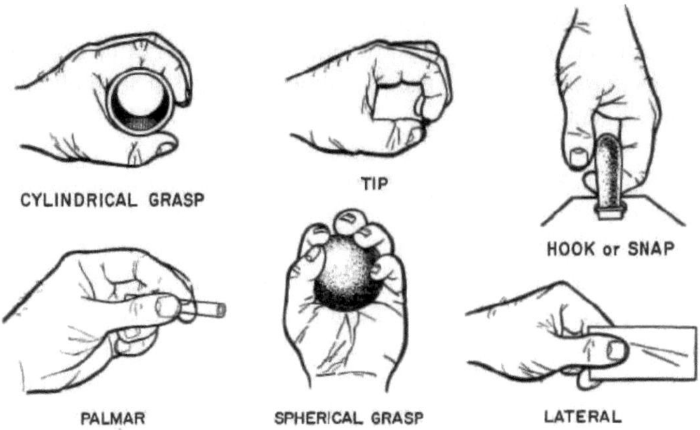

Fig. 3.2 The six basic prehensile patterns defined in [164], adapted from [179]

3.3 Grasp Preshapes and Shape Primitives

The concept of *grasp preshapes*, also called *hand preshapes*, *hand postures*, or *prehensile patterns*, was firstly studied by Schlesinger, in Germany, who proposed a simple grasp taxonomy for classifying the prehensile functionalities of prosthetic hands from an anatomical point of view [164]. This taxonomy was later adopted in [179] in a study of the structure of the human hand and wrist in terms of bones, joints and muscles. Based on the original Schlesinger's taxonomy, Taylor reported six different prehension patterns, shown in Figure 3.2, which are: cylindrical, tip, hook, palmar, spherical and lateral.

A prehensile pattern is a hand configuration useful for a grasp on a particular shape and for a given task. The Schlesinger's classification was based primarily on the object shape, without considering the task to be performed with the object. It was in [128] where the intended task was firstly considered as key factor for selecting a suitable hand posture, together with the shape of the object, its size and miscellaneous factors such as weight and temperature. In the Napier's approach, the task was represented as a tradeoff between *stability* and *precision*. He defined two distinct patterns providing the anatomical background for all prehensile movements:

- Power grasps, distinguished by large contacts of the fingers and the palm surface with the object. Power grasps are suitable for tasks requiring stability, at the expense of little ability to move the fingers, i.e. lost of precision.
- Precision grasps, where contacts are performed with the tips of the fingers and thumb, thus allowing to impart motion with the fingers. These grasps are suitable when the task requires increased sensitivity and dexterity.

These concepts were later adopted by the mechanics and robotics communities, the former for designing versatile robotic hands, and the latter as a useful way of reducing the complexity of contact-level grasp planning for a dexterous robotic hand. The authors in [105] developed a grasp index table for selecting the most suitable hand posture according to the desired grasp functionality (stability vs. precision), the object size and the shape. He extended Napier's taxonomy with an additional prehensile pattern which functionality was in the middle of the power and precision grasps: the *lateral pinch*, characterized for making contact with the whole surface of the index and thumb fingers in opposition (note that this pattern was already considered in the Schlesinger's taxonomy shown in Figure 3.2). As the contact surface is greater, the grasp offers more stability than precision grasps. Some finger motion is also possible, thus allowing more dexterity than power grasps. Lyons was the first in applying a grasp taxonomy for the high-level control of a robotic dexterous hand.

Cutkosky started from the original power and precision patterns of Napier's classification and developed a very complete taxonomy [33], designed to codify the knowledge required for manipulation tasks in a manufacturing environment. The authors studied the grasps used by humans in machining environments, with the goal to understand the relationship between the task requirements and the adopted grasp. In the Cutkosky's taxonomy, several prehensile patterns were classified according to its task-related and object-related properties. This taxonomy was the first providing a quite complete systematic mapping between the task requirements on a given object shape and an appropriate grasp. The most important preshapes, as described in [33] are: cylindrical grasp, power grasp, modified cylindrical grasp, spherical power grasp, modified hook grasp, five-fingertip grasp, four-finger precision grip, modified precision grip, three-finger precision grip, two-finger precision grip, lateral pinch and complex grip. Their study also suggested general principles for the design and control of manufacturing hands.

Since the publication of the Cutkosky's taxonomy, several researchers in the robotics community have adopted the grasp preshapes as a method for efficient and practical grasp planning in contrast to contact-based techniques. This new approach has received the name of *knowledge-based approach* to grasping [175]. From the robotics point of view, a grasp preshape is a set of finger postures adopted as the wrist moves towards the object. The grasp is generally performed by moving the wrist to a suitable position close to the object, and then closing the fingers until contact is made.

A *shape primitive* is a simple geometry used to approximate an object shape, and over which a grasp preshape can be easily planned. A shape primitive, together with the intended task, determines a prehensile pattern. This idea was already introduced in [175] and [9], where objects were modeled by elliptical cylinders, over which power grasps were easily computed. In reference [120] four different shape primitives were proposed, which could be combined for modeling any object geometry: spheres, cylinders, cones and boxes. A set of rules was used to compute an appropriate grasp preshape on the object, depending on its approximation by shape primitives. The hand posture was selected from a set of four grasp preshapes previously defined for the Barrett Hand. The authors developed a grasp planner following these ideas, which was later used by other authors for planning preshapes for a humanoid hand [121], or grasp planning on box-based object shape approximations [74], amongst others.

Grasp preshapes and shape primitives have shown to be an intuitive, efficient, practical and powerful approach to grasp planning, in contrast to contact-based approaches which rarely consider hand kinematics and practical viability. In addition, this approach is strongly based on anatomical studies of the human hand prehension functionalities, from the beginning of the 20th century until nowadays. For all of these reasons, the part of our research devoted to grasp planning and control is based on the knowledge-based approach.

3.4 A Framework for Physical Interaction: The Grasp Meets the Task (and Vice Versa)

The TFF and the knowledge-based approach to grasping are tools that have been used independently by the task planning and grasp planning communities, but the relationship between them has been never considered. A joint framework linking both approaches would allow an integrated specification of the grasp and the task, thus supporting a wide range of actions far beyond the generally adopted pick and place tasks.

Based on these well-established theories, we develop a framework for the specification and robust control of physical interaction tasks, where the grasp and the task are jointly considered on the basis of multi-sensor information, leading to task-oriented grasps and grasp-oriented tasks (hereon referred simply as *grasp* and *task* for the shake of readability). The grasp is defined as a desired relationship between the robot hand, shaped with a suitable prehensile posture, and the object to be manipulated. The task is defined as a desired constrained motion that must be applied to the object in terms of a TF specification.

In order to fill the gap between the grasp and the task, two auxiliary frames are defined, in addition to those already used by the underlying techniques and particular control approach: a *hand frame*, associated to a hand posture, and a *grasp frame*, associated to the object to be manipulated. On the one hand, the grasp frame is used as the goal for the grasping action. On the other hand, it is related to the task, through the object kinematic and geometric model. The task is then performed by transforming the desired constrained task motion, given in the TF, into a suitable cartesian robot motion, given in the robot end-effector coordinates, passing through the particular grasp configuration (the *hand-grasp link*). The hand and the grasp frame establish a link for transforming the task specification into robot motion.

As the particular grasp performed on the object can be subject to important uncertainties and disturbances during execution, information coming from multiple sensors must be used in order to compute an estimation of the hand-grasp relative pose. Having a suitable sensor-based estimation of this transformation, the task references, given in the TF, can be transformed into robot coordinates during execution, following a sensor-based TF tracking approach [19].

In this section, the TFF and the knowledge-based approach to grasping are linked through a set of *physical interaction frames*. A physical interaction task is then

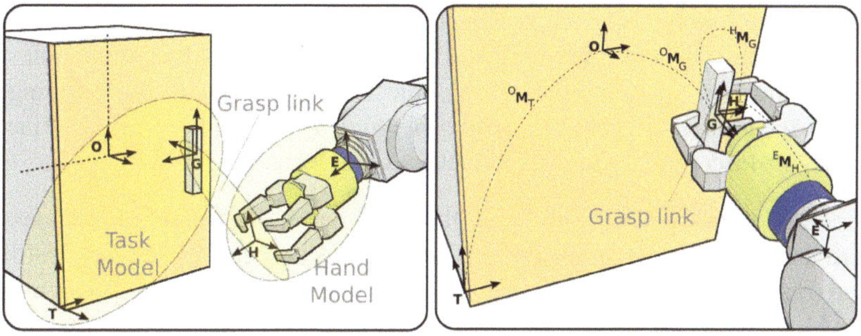

Fig. 3.3 The physical interaction frames are (left): the object frame (O), the end-effector frame (E), the task frame (T), the hand frame (H) and the grasp frame (G). The task motion must be transformed into robot coordinates through the kinematic chain formed by T, G, H and E (right). The *grasp link* is the relative pose between frames H and G, and represents the bridge between the task and the grasp.

described by a suitable placement of these frames, and the relationships between them. We study how sensors can be used for tracking the physical interaction frames, and show some conceptual examples of daily physical interaction tasks specified with the proposed approach.

3.4.1 The Physical Interaction Frames

In order to assist the specification of a physical interaction task, five auxiliary frames are used, as shown in Figure 3.3:

The *Object frame (O)*, which is the origin where the object is defined.

The *End-effector frame (E)*, where the control of the robot is performed. We focus on cartesian control of the end-effector, because it has a direct mapping with the constrained motion, which is also specified in the cartesian space. An inverse kinematics controller is needed in order to transform cartesian position/velocities into a suitable motion of the arm and the mobile platform.

The *Task frame (T)*, where the task is specified according to the TFF. The programmer, or task planner, has to choose a suitable task frame according to the requirements originally described in [19].

The *Hand frame (H)*, which is a coordinate system attached to the robot hand (or tool). It depends on the adopted hand posture and control strategy used for making contact. For control purposes, it is necessary to link the hand frame with the robot end-effector frame. This can be normally done through robot hand kinematics. In an analogy with the feature frames defined in [37], the hand frame can be placed on a *physical entity*, like the fingertip surface, or on an *abstract entity*, like the middle point in the imaginary line joining the thumb and the index fingers.

The *Grasp frame (G)*, given in object coordinates, and related to the *task frame* through the object geometry, or through an user-defined transformation. This frame must be set to the part of the object which is suitable for grasping and task execution. It can also be placed on a *physical entity*, like a button surface, or on an *abstract entity*, like the symmetry axis of a handle.

Figure 3.3 shows an illustration of the different frames and its relationships, before and after the grasping action. The relationship between the physical interaction frames is represented with the following homogeneous transformation matrices: ${}^{E}\mathbf{M}_{H}$, ${}^{O}\mathbf{M}_{G}$, ${}^{O}\mathbf{M}_{T}$ and ${}^{H}\mathbf{M}_{G}$, relating, respectively, the end-effector frame to the hand frame, the object frame to the grasp frame, the object frame to the task frame and the hand frame to the grasp frame. Each transformation is composed of a rotation matrix and a translation vector, i.e. ${}^{i}\mathbf{M}_{j} = \begin{bmatrix} {}^{i}\mathbf{R}_{j} & {}^{i}\mathbf{t}_{j} \\ \mathbf{0}_{1\times 3} & 1 \end{bmatrix}$, where ${}^{i}\mathbf{R}_{j}$ is the 3×3 rotation matrix between frames i and j, and ${}^{i}\mathbf{t}_{j}$ represents the position of frame j with respect to frame i.

The object, task and grasp frames compose the *task model*, whereas the end-effector and hand frames establish the *hand model*. It is worth noting that the specification of a task model does not necessarily require an object geometric model. The homogeneous transformations ${}^{O}\mathbf{M}_{T}$ and ${}^{O}\mathbf{M}_{G}$, can be set from an object model, but also from a user-defined specification not involving a geometric model, or from robot sensors, like in the interactive perception approach of [86]. The same concept applies to the hand model: ${}^{E}\mathbf{M}_{H}$ can be computed from hand kinematics, but also from visual feedback or a user-defined transformation. Thus, the framework is defined in an object-independent manner. The physical interaction frames can be set with or without using object geometric models.

The task execution requires computing which motion is necessary at the end-effector frame in order to achieve a desired force or velocity at the task frame. For this, the velocity and force references given in the TF must be expressed into robot-centered coordinates through the kinematic chain composed of the frames T, G, H and E. However, the relative pose between the robot hand and the part of the object being manipulated, ${}^{H}\mathbf{M}_{G}$, can vary during the particular execution and, in general, is subject to important geometric uncertainties. This transformation matrix represents the link for transforming the task description into robot coordinates and must be continuously estimated by the robot sensors. We will refer to it indistinctly as the *grasp link*, the *hand-grasp link*, *hand-grasp transformation* or the *hand-grasp relationship*.

3.4.2 A Physical Interaction Task

A physical interaction task is composed of the grasping action and the subsequent possibly constrained interaction with the environment. Both the grasp and the task specification are detailed in the following sections, based on the physical interaction frames defined previously. Some conceptual examples are also provided.

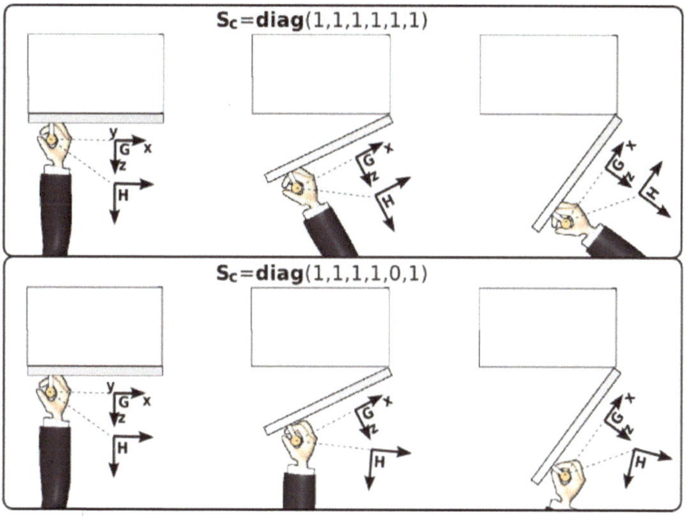

Fig. 3.4 A full-constrained grasp (top row) vs. an under-constrained grasp (bottom row). An under-constrained grasp is characterized because the grasp and the hand frames are not rigidly attached during the task execution.

3.4.2.1 The Grasp

The grasping action is specified as moving the hand frame towards a desired relative positioning with respect to the grasp frame, i.e., taking the grasp link towards a desired configuration.

Constrained and free degrees of freedom for the grasp can be indicated. For the constrained DOF's, the hand frame must completely reach the desired relative pose with respect to the grasp frame. However, for the free DOF's, there is no particular relative pose used as reference. Instead, the robot controller can choose a suitable pose, according to different criteria such as manipulability, joint limit avoidance, etc. We refer to these directions as *grasp-redundant DOF's*. A grasp with grasp-redundant DOF's is called an *under-constrained grasp*. An example of an under-constrained grasp is shown in Figure 3.4: the rotation around the handle axis is a grasp-redundant DOF that can be used by the controller in order to achieve a secondary task. As we will see in chapter 5, the control of grasp redundancy can greatly expand the robot workspace during physical interaction.

Let $\mathscr{P} = \{m_0, m_1, \ldots, m_n\}$ represent a hand preshape (either prehensile or non-prehensile), where m_i is the desired value for each of the n DOF's of the hand. The grasp is then defined as:

$$\mathscr{G} = \left\{\mathscr{P}, H, G, {}^{H}\mathbf{M}_{G}^{*}, \mathbf{S_c}\right\} \tag{3.2}$$

where ${}^{H}\mathbf{M}_G^*$ is an homogeneous transformation matrix that contains the desired relationship between the hand and the grasp frame, whereas $\mathbf{S_c}$ is a 6×6 diagonal selection matrix, given in the grasp frame, which indicates the cartesian degrees of freedom constrained by the grasp. A value of 1 at the diagonal element i indicates that the corresponding DOF is constrained by the grasp, whereas a value of 0 indicates that it is not. Therefore, the grasp is specified as a desired relative pose (possibly under-constrained) between the hand frame and the grasp frame, i.e. a desired state of the grasp link.

3.4.2.2 The Task

The task requires performing compliant motion, following a set of velocity-force references defined in the task frame, according to the TFF. It is defined as follows:

$$\mathscr{T} = \{T, \mathbf{v}_T^*, \mathbf{f}_T^*, \mathbf{S_f}\} \tag{3.3}$$

$\mathbf{S_f}$ is a 6×6 diagonal selection matrix, where a value of 1 at the diagonal element i indicates that the corresponding DOF is controlled with a force reference, whereas a value of 0 indicates it is controlled with a velocity reference. This matrix is equivalent to the *compliant selection matrix* of [152]. A velocity reference is suitable for tasks where a desired motion is expected, whereas a force reference is preferred for dynamic interaction with the environment, where no object motion is expected, but a force must be applied (for polishing a surface, for example). $\mathbf{v}_T^*$ and $\mathbf{f}_T^*$ are, respectively, the velocity and force reference vectors given in the task frame, T.

3.4.2.3 Examples

Figure 3.5 shows three daily physical interaction situations that can be specified with the proposed framework. The first is an example of a task where a dynamic interaction with the environment is desired. Instead of specifying a velocity, the task is described as a desired force to apply to a button, along Z axis of the task frame T. The hand frame, H, is set to the fingertip of a *one-finger preshape*. The grasp frame, G, is set to the button surface so that the desired hand-grasp transformation can be set to the identity, which corresponds to the case where the fingertip is in contact with the button. This is the most common example of a non-prehensile grasp. For this case, the robot may choose the most suitable relative rotation around Z axis of the grasp-hand frame. Thus, Z axis is set to be a free DOF.

In the second example, a rotation velocity around Z axis of the task frame, T, is desired in order to turn on the tap. The grasp frame, G, is set to a part suitable for grasping, whereas the hand frame is set to the middle point between the thumb and the index finger in a *pinch preshape*. For performing the grasp, the hand frame must match with the grasp frame, up to a rotation about Y axis, which is set to be a grasp-redundant DOF.

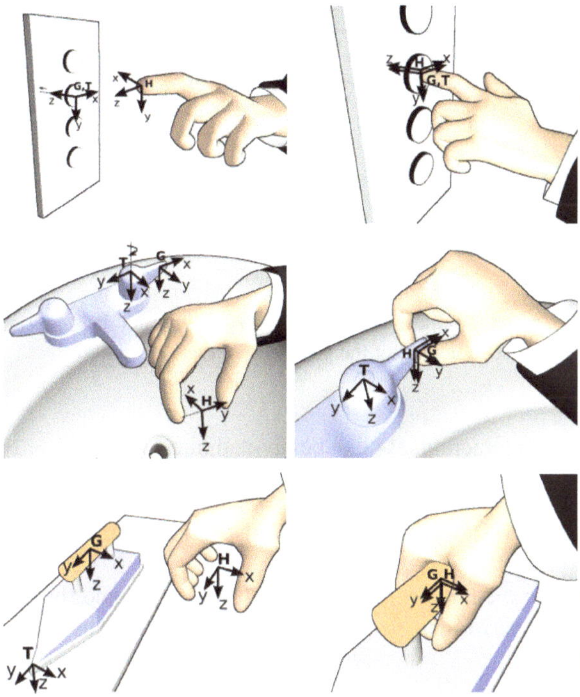

$\mathscr{P}$ = "*one-finger preshape*"
$^{H}\mathbf{M}_{G}^{*} = \mathbf{I}_{4\times4}$
$\mathbf{S_c} = \mathbf{diag}(1,1,1,1,1,0)$
$\mathbf{S_f} = \mathbf{diag}(0,0,1,0,0,0)$
$\mathbf{v}_T^{*} = (0,0,0,0,0,0)$
$\mathbf{f}_T^{*} = (0,0,f_z^{*},0,0,0)$

$\mathscr{P}$ = "*pinch preshape*"
$^{H}\mathbf{M}_{G}^{*} = \mathbf{I}_{4\times4}$
$\mathbf{S_c} = \mathbf{diag}(1,1,1,1,0,1)$
$\mathbf{S_f} = \mathbf{diag}(0,0,0,0,0,0)$
$\mathbf{v}_T^{*} = (0,0,0,0,0,w_z^{*})$
$\mathbf{f}_T^{*} = (0,0,0,0,0,0)$

$\mathscr{P}$ = "*cylindrical power*"
$^{H}\mathbf{M}_{G}^{*} = \mathbf{I}_{4\times4}$
$\mathbf{S_c} = \mathbf{diag}(1,1,1,1,0,1)$
$\mathbf{S_f} = \mathbf{diag}(0,0,1,0,0,0)$
$\mathbf{v}_T^{*} = \mathbf{f}(t)$
$\mathbf{f}_T^{*} = (0,0,f_z^{*},0,0,0)$

Fig. 3.5 Some physical interaction situations described with the framework. First: pushing a button, with a force reference. Second: turning on a tap, with a velocity reference. Third: ironing task, with a velocity and force reference.

Finally, the third example shows an ironing task where both a velocity and a force reference are needed. The Z axis of the task frame is force-controlled in order to make a reference force against the ironing board. At the same time, axis X and Y are velocity-controlled in order to follow a particular trajectory, $\mathbf{f}(t)$. Regarding the grasp, a *cylindrical power preshape* is adopted, with a free DOF around Y axis of the hand frame, H.

In all of the cases, the velocity and force references must be set to suitable values. When applying this framework to real autonomous robots, these references should be set automatically by a planner, depending on the particular case.

3.4.3 *Execution*

In the previous points we have defined the necessary elements for the specification of physical interaction tasks. In this section, we focus on their execution. However, instead of defining a detailed control law at this level, we rather describe some guidelines for the implementation of a suitable physical interaction controller.

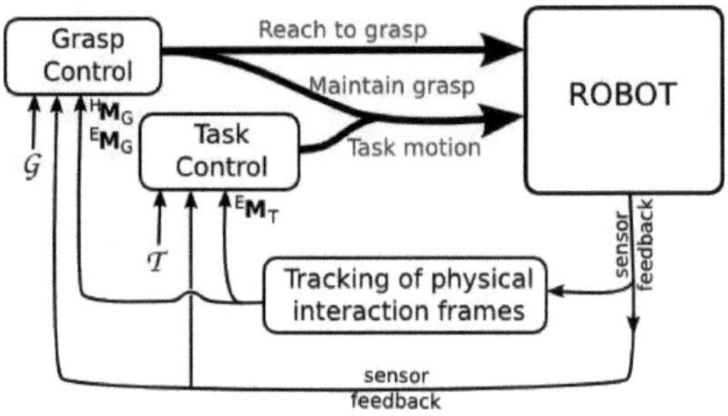

Fig. 3.6 The general control approach, composed of the *grasp controller* and the *task controller*, both relying on the sensor-based tracking of the physical interaction frames

The execution of a physical interaction task can be divided into two steps:

- A *non-contact* phase, corresponding to the reach to grasp action, where the hand of the robot must be moved towards the object until the task-oriented grasp is executed successfully.
- An *interaction* phase, where the hand is in contact with the object and the task constrained motion must be performed through robot motion, while keeping an appropriate contact situation.

During interaction, it is very important to maintain a good grasp on the object, specially for non-prehensile grasps which do not constrain all the relative hand-object motions. A suitable controller should perform the motion that ensures the execution of the task, but also auxiliary motion aimed at constantly improving, or at least maintaining, the grasp link.

We propose a general control scheme, composed of two simultaneous controllers, as shown in Figure 3.6: the *grasp controller* and the *task controller*. The grasp controller is in charge of the reaching to grasp action, but also of the auxiliary motion required for maintaining an appropriate grasp condition during the task motion. Regarding the task controller, it is involved in the execution of the compliant motion required for the task. The consideration of the grasp state during the task execution represents an important aspect of our approach to grasp-task integration, and allows to increase the execution reliability, as hand-object misalignments can be detected and corrected. Both the grasp and task controllers rely on sensor information about the location of the physical interaction frames. See low-level details in the following.

3.4.3.1 Grasp Control

The grasp controller must ensure that the desired relationship between the hand and the grasp frame is achieved, at the same time that fingers are controlled to reach the desired hand posture. Using sensor feedback and an estimation of the state of the grasp link, the grasp controller generates control signals for the arm and for the hand

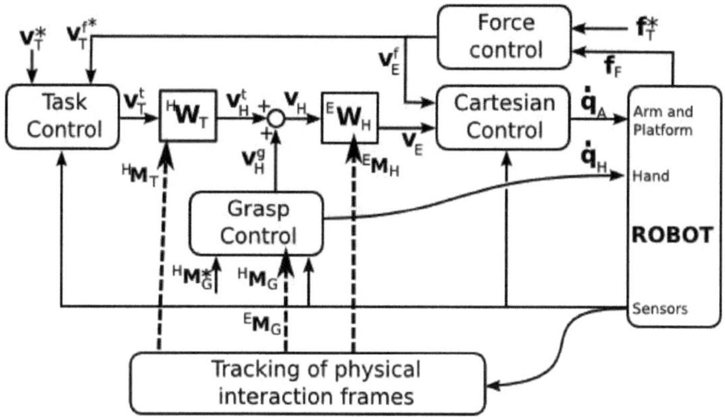

Fig. 3.7 A more detailed view of the general control approach. The grasp controller generates control signals for the arm and the hand, based on sensor feedback and the estimated pose of the grasp frame with respect to the hand frame. The task controller generates arm control signals for performing the compliant motion, based on the sensor-based tracking of the task frame and force control.

(see Figure 3.7). The arm motion is in charge of approaching the hand to the target object, whereas the finger motion actually performs the preshaping of the hand.

For the arm motion, the simplest case is to design a proportional controller moving the hand in a straight line towards the target. Let ${}^{H}\mathbf{M}_{H}^{*} = {}^{H}\mathbf{M}_{G} \cdot \left({}^{H}\mathbf{M}_{G}^{*}\right)^{-1}$. Then, a proportional position-based control can be performed with the following equation, where $\mathbf{h}_{H}^{*}$ is a pose vector build from the homogeneous matrix ${}^{H}\mathbf{M}_{H}^{*}$, λ_p is the control gain, and $\mathbf{v}_{H}^{g}$ is the resulting velocity of the grasp controller, given in the hand frame:

$$\mathbf{v}_{H}^{g} = \lambda_p \mathbf{h}_{H}^{*} \tag{3.4}$$

When the hand is far from the target, this controller is the simplest case of reaching. It can be done in open loop if a good estimation of the target pose with respect to the hand is available [146], but closed loop is more adequate if we want to deal with the uncertainties of non-structured environments in a sensor-based approach, as it will be shown in the practical examples of chapter 6. In closed loop, the grasp link, ${}^{H}\mathbf{M}_{G}$, must be computed at each control cycle from the robot sensor feedback.

During task execution, the grasp controller can correct the relative misalignments between the hand and the object, with the purpose of keeping a suitable grasp. Misalignments can appear due to errors in the geometric models, robot calibration, or a TF specification violating the *Geometric Compatibility* requirement, for example. It is of utmost importance to keep an appropriate hand-grasp relationship during interaction, because the successful execution of the task depends on it.

Other more advanced control solutions are possible. In an environment where obstacles are modeled, a path planning algorithm could provide a feasible collision-free path, not necessarily in a straight line. In addition, for those objects requiring special grasping procedures, it would be possible to implement an event-based grasping approach, where sensor feedback is not only used inside the control loop, but also for triggering conditions that allow to switch between different atomic actions (an example will be described in chapter 5). It is also worth noting that the control equation 3.4 does not consider grasp redundancy, whereas it may be necessary for the success of some tasks. Our purpose in this section is only to present some general guidelines concerning implementation of the grasp controller. Chapters 5, 6 and 7 will show specific controllers, based on the general control scheme proposed here, for different tasks and sensor combinations.

3.4.3.2 Task Control

The task controller must be able to transform a velocity-force reference, $\mathbf{v}_T^*$ and $\mathbf{f}_T^*$, given in the task frame, into a hand velocity, $\mathbf{v}_H^t$, and finally into joint motion, as shown in Figure 3.7. For this, the task velocity-force reference must be transformed into a unique velocity reference given in the task frame, $\mathbf{v}_T^t$. The TF directions which are velocity-controlled can take directly the velocity reference, $\mathbf{v}_T^*$. However, those which are force-controlled need to transform an explicit force reference, $\mathbf{f}_T^*$, into a position or velocity setpoint.

Several explicit force control methods have been proposed in the literature, and some of them will be outlined in chapter 5. If $\mathbf{v}_T^{f^*}$ is a *force-based* velocity computed by an explicit force control approach, then the task velocity can be computed as:

$$\mathbf{v}_T^t = \mathbf{S_f}\mathbf{v}_T^{f^*} + (\mathbf{I} - \mathbf{S_f})\mathbf{v}_T^* \tag{3.5}$$

The task velocity, $\mathbf{v}_T^t$, must be later transformed into another velocity reference given in the frame where the cartesian control of the robot is performed, i.e. the end-effector frame. This can be done by following the kinematic chain composed of the task, grasp, hand and end-effector frames:

$$\mathbf{v}_E = \underbrace{{}^E\mathbf{W}_H \cdot {}^H\mathbf{W}_G \cdot {}^G\mathbf{W}_T}_{{}^E\mathbf{W}_T} \cdot \mathbf{v}_T^t \tag{3.6}$$

where ${}^i\mathbf{W}_j$ is the 6×6 twist transformation matrix associated to ${}^i\mathbf{M}_j$, and defined as:

$${}^i\mathbf{W}_j = \begin{pmatrix} {}^i\mathbf{R}_j & [{}^i\mathbf{t}_j]_\times {}^i\mathbf{R}_j \\ 0 & {}^i\mathbf{R}_j \end{pmatrix} \tag{3.7}$$

$$[\mathbf{v}]_\times = \begin{pmatrix} 0 & -v_z & v_y \\ v_z & 0 & -v_x \\ -v_y & v_x & 0 \end{pmatrix} \tag{3.8}$$

Finally, the end-effector velocity must be transformed, by a low-level controller, from the cartesian space into joint space through the mobile manipulator cartesian controller. The implementation of cartesian controllers is out of the scope of this book. Only chapter 5 will describe the control of humanoid arm in the operational space, with the purpose to show how joint and grasp redundancy can be used in real applications. On the rest of practical implementations, we will assume that a cartesian controller is already available.

During the interaction phase, the robot hand is in contact with the environment, and any kind of uncertainty may produce considerable forces that can damage the environment or the robot. When the robot is in contact with the environment, it is extremely important to design a controller that can deal with unpredicted forces and adapt the hand motion accordingly. Thus, force-torque feedback is a necessary condition (and, in some cases, sufficient) for robust manipulation. Whereas some tasks can still be performed when deprived of visual and tactile feedback, force feedback cannot be removed if success in manipulation is desired. Thus, we assume that a force-torque sensor is always present, so that force controllers can be implemented in order to deal with the undesired forces.

Some specific examples following these guidelines will be shown in chapters 5, 6 and 7.

3.4.3.3 Sensor-Based Tracking of the Physical Interaction Frames

The general grasp and task controllers described in the previous points, depend completely on the relative pose between the physical interaction frames. The grasp controller needs to know the grasp link, ${}^{H}\mathbf{M}_{G}$, which is also required by the task controller, as part of the transformation ${}^{E}\mathbf{M}_{T}$ that links the task with the robot end-effector, i.e. ${}^{E}\mathbf{M}_{T} = {}^{E}\mathbf{M}_{H} \cdot {}^{H}\mathbf{M}_{G} \cdot {}^{G}\mathbf{M}_{T}$.

We support the idea that robot perception must continuously provide information about the relative pose between the physical interaction frames. In most of the cases, ${}^{E}\mathbf{M}_{H}$ and ${}^{G}\mathbf{M}_{T}$ can be computed from hand kinematics and the task model respectively. One exception, considered in detail in the next section, is the use of tools. In this case, the hand frame can be placed on the tool so that its position with respect to the end-effector cannot be computed directly from the hand encoders. Additional sensors would be needed in order to estimate the ${}^{E}\mathbf{M}_{H}$ transformation in this case.

However, the grasp link transformation, ${}^{H}\mathbf{M}_{G}$, must be definitively estimated through robot sensors, because it can be subject to important execution uncertainties such as bad positioning, poor sensory information, sliding, etc. The robot must always estimate the hand-grasp transformation during task execution in order to assist the grasp and task controllers. This transformation is necessary for tracking the task frame position and orientation during interaction. The estimation of ${}^{H}\mathbf{M}_{G}$ is the key for relating the task frame to robot coordinates, thus allowing the transformation of the task constraints into suitable robot motion (see Figure 3.8).

The best sensor input to estimate this relationship is vision. A robot could be observing its hand and the object simultaneously, while applying model-based pose estimation techniques [44, 28]. Another interesting sensor is a tactile array, which provides detailed local information about contact, and could be used to detect grasp

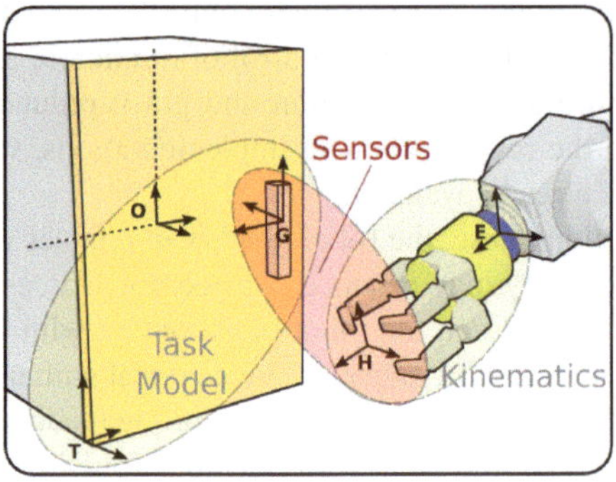

Fig. 3.8 The use of sensor feedback is specially important for the estimation of the grasp link. Whereas the ${}^{E}\mathbf{M}_H$ and ${}^{G}\mathbf{M}_T$ transformations can be computed from hand kinematics and the object model in most of the cases, ${}^{H}\mathbf{M}_G$ can only be estimated through sensors if the environment is not completely modeled.

mistakes or misalignments. In general, the best solution is to combine several sensor modalities for getting a robust estimation. Chapters 5, 6 and 7 will describe several applications combining vision, force and tactile sensors with the purpose of monitoring the grasp link.

3.4.4 Use of Tools

The proposed framework also supports the specification of physical interaction tasks involving tools. We can consider the tool as an extension to the hand, so that the hand frame can also be set to parts of the tool. However, in this case, the relationship between the end-effector frame and the hand frame cannot be computed only through hand kinematics. It depends on the exact position of the tool in the hand, and must be captured by sensors once the utensil has been grasped. Tactile and vision integration can play an important role here.

The use of tools consists of two different instances of physical interaction tasks: one, involving a grasp for a transport task, and another one, involving the actual use of the tool.

In order to bring the tool to the object, the robot must first grasp the tool and pick it up. This task can be easily described with the proposed framework, by choosing a suitable hand preshape and grasp frame depending on the particular tool and the task to perform, as shown in Figure 3.9a.

Once the tool is grasped, it can be considered as an extension to the robot hand, and the proposed formalism still applies. Using the tool requires moving the tool towards the object (reaching) and performing the task motion. During the reaching phase, the hand frame can be set to a part of the tool suitable for making contact with

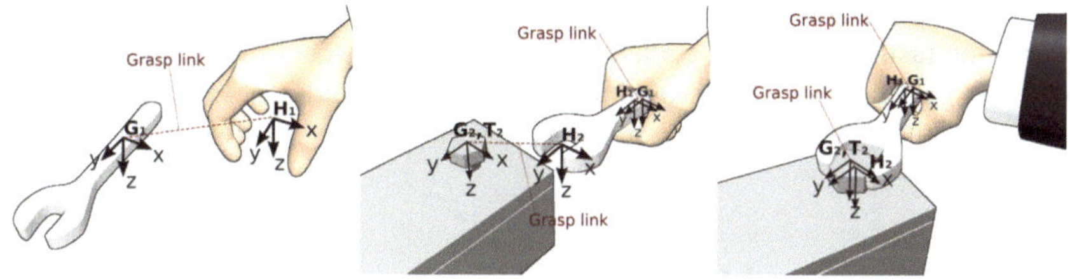

(a) Grasping the tool (b) Reaching for the object (c) Using the tool

Fig. 3.9 An example of tool grasping and use based on the proposed formalism

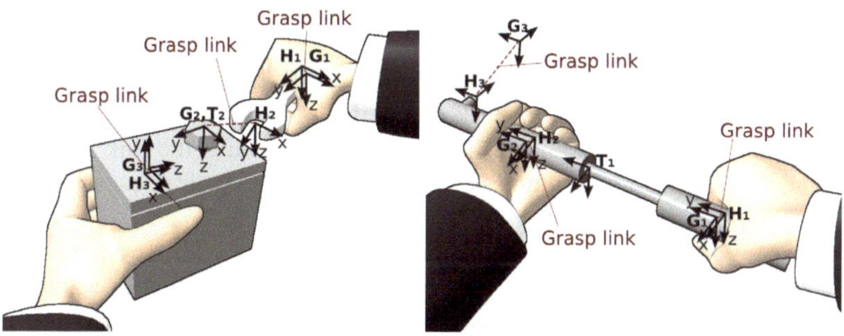

(a) One hand is holding an object while the hand uses a tool on it. (b) One hand holds a bicycle pump while the other performs the inflating task.

Fig. 3.10 Two examples of two-handed manipulation under the proposed framework

the object. Therefore, reaching can be defined in terms of the proposed formalism, as an alignment between the hand frame, placed on the tool in this case, and the grasp frame, as shown in Figure 3.9b,c.

When the tool is aligned with the object part, the task can be transformed from object coordinates to tool coordinates, and finally to end-effector coordinates, passing through two different grasp links. Sensor feedback must contribute in order to compute an estimation of the relative pose between, first, the object and the tool, and second, the tool and the hand. Therefore, the use of tools requires to apply the proposed formalism simultaneously for two different subsystems: hand-tool and tool-object.

3.4.5 Two-Hand Manipulation

In the case of physical interaction with two hands, we can even have three different subsystems running simultaneously under the proposed framework. For example, one hand could be holding an object, whereas the other hand could be holding a tool and acting on the object, as shown in Figure 3.10a. In this case, one physical interaction task would be needed for one hand (hand-object), whereas two physical interaction tasks would be specified for the other hand (hand-tool and tool-object). Another situation could be an inflating task, where a hand is holding a tool acting on an object, whereas the other hand works on the tool (see Figure 3.10b).

3.5 Discussion and Conclusions

In this chapter, a framework for the integrated specification of the grasp and the task has been proposed, based on well-established techniques adopted from the task control and grasping communities: the task frame formalism, and the knowledge-based approach to grasping.

The framework is based on the specification of five different frames: the object frame, the end-effector frame, the task frame, the hand frame and the grasp frame. Among them, the hand frame and the grasp frame establish the link between the grasp and the task. Several examples of the application of this framework have been presented.

The execution of the physical interaction task is performed by two simultaneous controllers: the grasp and the task controller. The grasp controller drives the hand towards a desired relative configuration with respect to the part of the object to be grasped, and tries to keep it during the task. The task controller transforms the task specification from the task frame to a set of joint velocities, passing through the hand-grasp link.

The control framework assumes that an estimation of the hand-grasp relationship exists. This relationship must be estimated during execution in order to easily transform the task specification, from object coordinates to robot coordinates. It can be obtained from the object model, robot hand kinematics and, specially, sensors.

More complex tasks, such as using tools or two-handed manipulation, can be described as a set of individual physical interaction tasks and, thus, specified through different instantiations of this framework.

This chapter has presented the general framework underlying the rest of this book. The proposed approach constitutes a versatile solution to the new problem of integrated grasp-task specification, as it allows to represent many different actions, not only involving grasping, but also dynamic interaction with the environment, use of tools and two-hand manipulation. Next chapters are focused on autonomy and dependability. More concretely, chapter 4 describes an approach for the autonomous planning of physical interaction tasks, whereas chapters 5, 6 and 7 concern the dependable execution of these tasks by means of force, vision and tactile control.

Chapter 4
Planning of Physical Interaction Tasks

4.1 Brief Introduction

In the previous chapter, a framework for the joint specification of the grasp and the task has been proposed. The main idea is that a physical interaction task can be described through a suitable specification of the physical interaction frames and the relationship between them. The proposed framework is based on well-established approaches to grasp and task planning, and part of the physical interaction frames are adopted from the underlying techniques.

If the particular physical interaction task that the robot must perform is known beforehand, a programmer can apply the previous framework for the manual specification and implementation of the grasping task and the compliant interaction with the environment. This is the typical case in a structured environment, where the range of possible tasks is limited and the robot programmer can implement a solution case by case.

In contrast, an unstructured scenario contains many possibilities, and it is not possible to program in advance all the physical interaction tasks that a robot may encounter. The robot must deal with many types of objects, handles, buttons, etc. in a versatile and autonomous manner. In the ideal case, a robot should be able to interact with objects never seen before. From a given high-level task description, the robot must be able to plan autonomously the physical interaction task, including the grasping action and subsequent compliant interaction, even for new objects. It must be able to adapt to the particular situation without being specifically programmed for it.

In this chapter, we describe a physical interaction planner, based on the framework for physical interaction, that allows to build automatically a task specification, starting from a high-level task description. We focus on interaction in home environments, which includes manipulation of furniture and home appliances: opening doors and drawers, switching lights, turning knobs, etc.

In our way to describe the planner, the concept of *task-oriented hand preshapes* is introduced in section 4.2, as an extension to the classical understanding of hand

M. Prats et al.: Robot Physical Interaction through the Combination of Vision, STAR 84, pp. 39–53.
springerlink.com

preshapes. We define a set of *ideal task-oriented hand preshapes* for an *ideal hand*, and then propose the *hand adaptors*, as a method for the instantiation of the ideal preshapes on real robotic hands. The main advantage of this approach is that the same planning algorithms can be used for different hands, just by defining a suitable mapping between the ideal hand and the real one, thus gaining in versatility.

In order to assist automatic planning, in section 4.3 we propose a box-based object representation where each object part can be classified into one category, indicating all the possible *object actions* that can be performed with it. An automatic method for labelling an object part, according to its geometry and kinematics, is also presented. An *object task* is then described in section 4.4 as a sequence of one or more object actions. Finally, in section 4.5 we show how a physical interaction task can be automatically specified, taking as input an object description and the action to perform on a given object part.

It is worth noting that the planner proposed in this chapter only considers a limited set of tasks concerning interaction with articulated parts in home environments, and considering very basic geometry and a limited set of grasp preshapes. Other physical interaction planners, based on the framework of the previous chapter, and taking into account other tasks or more complex geometry, would be possible,

4.2 Task-Oriented Hand Preshapes

The task is a very important factor to have into account when adopting a grasp preshape. In the study of prehensile patterns of Napier [128], the task played a very important role when deciding which pattern to apply. Cutkosky & Wright [33] also focused their grasp taxonomy on the task requirements. However, most of the practical knowledge-based approaches to grasping do not consider any task other than pick and place.

In this work, the knowledge-based approach to grasping is adopted in a task-oriented manner. We extend the classical concept of hand preshape, as it is usually considered in robotics, for the inclusion of a task-oriented entity: the hand frame. As defined by the framework for physical interaction described in chapter 3, the hand frame is used to establish a link between the grasp and the task. Therefore, we introduce a task-oriented entity into the grasp preshape concept with the purpose of enabling knowledge-based task-oriented grasping.

We define a *task-oriented hand preshape* as a hand preshape extended with a hand frame. The hand frame is not only used for the grasp link specification, but also for describing the part of the hand that will be used for the physical interaction task. For example, pushing a button requires a contact with the fingertip, in contrast to a power grasp where contact is primarily done with the palm and the inner surface of the fingers. In the first case, the hand frame would be set to the fingertip, whereas in the second case it could be placed on the palm. One original hand preshape can generate several task-oriented hand preshapes, depending on the placement of the hand frame.

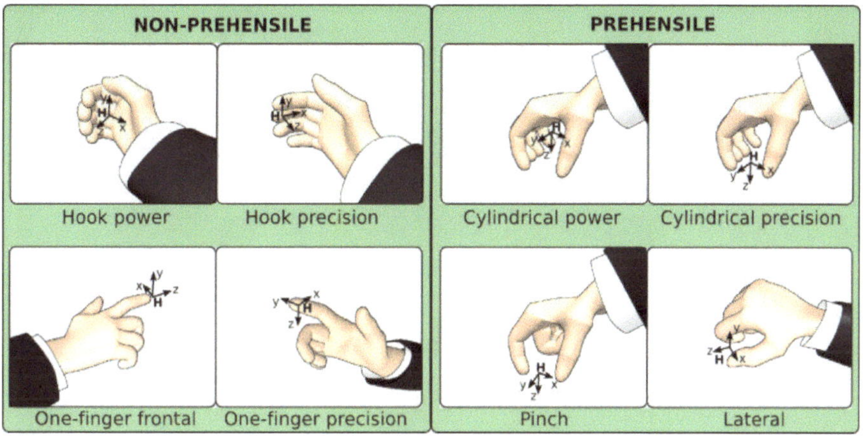

Fig. 4.1 The ideal task-oriented hand preshapes considered by the physical interaction task planner

We define a set of task-oriented hand preshapes for an *ideal hand*, which is an imaginary hand able to perform all the human hand movements. The task-oriented hand preshapes defined for this hand are called the *ideal task-oriented hand preshapes*, or *ideal hand preshapes* for simplification. Figure 4.1 shows the ideal task-oriented hand preshapes considered by the planner. They are:

Hook power. The hand adopts a hook posture. Contact is performed with the inner part of the fingertips, near the palm. The thumb finger does not contribute to the grasp. This grasp is useful when the object can be enveloped with the fingers, and the task requires making force along one or two directions (Z and/or X in the hand frame). One example is turning a handle and pulling back.
Hook precision. The hand adopts a hook posture, as in the previous case, but the hand frame is placed at the fingertips. This grasp is suitable for pushing along one direction (Z in the hand frame), when it is not possible to envelope the object with the fingers. Pushing objects, or opening a sliding door are two examples for which this grasp could be needed.
Cylindrical power. The hand takes a cylindrical configuration, where the thumb is in opposition to the rest of the fingers. The hand frame is placed on the center of the polyhedron generated by taking the fingertips and the palm as vertex. This preshape is useful for enveloping objects and applying forces in all the directions.
Cylindrical precision. The hand takes a cylindrical configuration, as in the previous case. The hand frame is placed on the centroid of the polygon generated by taking all the fingertips as vertex. This preshape is useful for grasping small objects and applying forces along the X direction of the hand frame.
One-finger frontal. This preshape is useful for pushing small objects like buttons. Only the index finger is used, with the hand frame placed at the fingertip. Forces can be applied mainly along the positive sense of the Z axis.

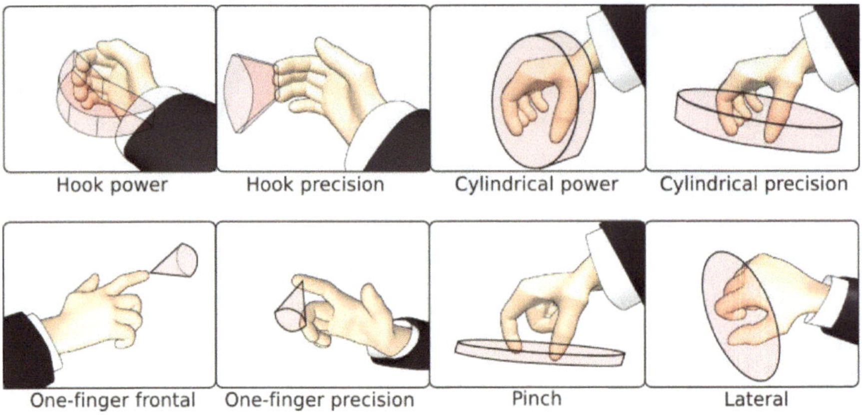

Fig. 4.2 An illustration of the translational components of the GWS that can be generated by each ideal hand preshape

One-finger precision. Similar to the previous case, but with the hand frame placed on the inner part of the fingertip. This preshape is useful for sensitive tasks, where tactile receptors at the fingertip play an important role. One example is grasping a book from a bookshelf, as we will see in chapter 8.

Pinch. The pinch preshape is similar to the cylindrical precision preshape, but only the index and thumb fingers are used. It is suitable for grasping small objects or interacting with controls attached to prismatic joints (such as a volume bar, for example).

Lateral. The grasp is performed with the thumb finger in opposition to the lateral part of the index finger. The preshape is appropriate for precision grasps that require applying considerable forces and torques, such as unscrewing a bottle cap, turning a knob, etc.

Each of these grasp preshapes is suitable for a particular set of tasks. Figure 4.2 shows a simplified visualization of the GWS that can be generated by each preshape. Precision preshapes are suitable for applying forces along well-known directions, in contrast to power preshapes, which can apply and resist forces in a wide range of directions. Therefore, if the interaction with the environment requires pushing along one known direction, *hook precision*, *cylindrical precision*, *pinch*, *one-finger frontal* and *one-finger precision* could be valid preshapes, depending on the object geometry. In contrast, interaction tasks requiring a wider force range would need a *hook power* or *cylindrical power* preshape. The *lateral preshape* is in the middle between precision and power, and should be used for interaction tasks that require a considerable force, when the object geometry does not allow a power grasp.

The physical interaction planner must choose the most suitable ideal preshape for the given object geometry and the intended task. However, for its application in a real robot, the ideal hand preshapes must be instantiated into real postures on a robotic hand, by means of *hand adaptors*. A hand adaptor is able to map an ideal preshape into a real robot preshape. The main advantage of planning on an ideal

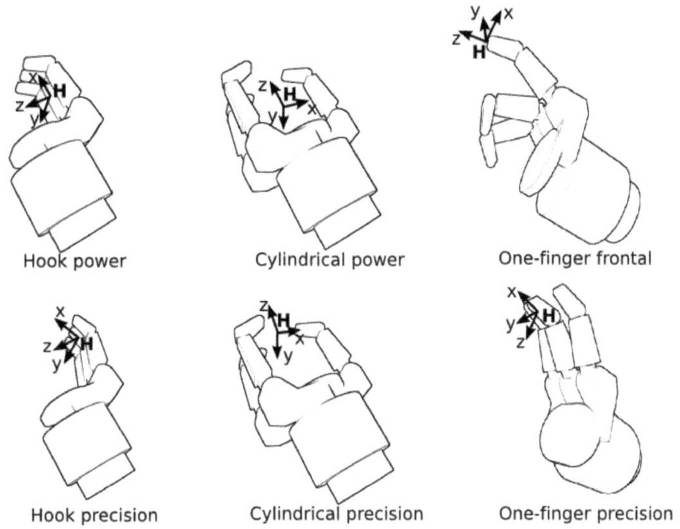

Fig. 4.3 The Barrett Hand task-oriented hand preshapes

hand and adapting the results to the real case is that the physical interaction planner can be independent of a particular robotic hand, thus providing versatility to the system. The result of the planner is general and can be easily applied in different robotic systems, just by defining the corresponding hand adaptor. The mapping for a Barrett Hand and a parallel jaw gripper are described in the following.

4.2.1 Mapping to the Barrett Hand

The Barrett Hand is a robotic hand that consists of three fingers and four motors. Three of the motors control the opening of each finger, whereas the fourth motor controls the *spread* mechanism, by which the external fingers can be placed in opposition to the middle one. The flexion of the outer link of each finger is coupled with the motion of the proximal link. A patented mechanism allows to perform the outer link flexion even if the proximal link is blocked by an obstacle.

As shown in Figure 4.3, six task-oriented hand preshapes are defined for the Barrett Hand, which are equivalent to the first six ideal preshapes:

Hook power. The three fingers are placed in a hook configuration, and the hand frame is placed on the inner part of the proximal link in the middle finger.
Hook precision. Equivalent to the previous preshape, but the hand frame is set to the inner part of the outer link.
Cylindrical power. The external fingers are placed in opposition to the middle finger. The hand frame is set to the centroid of the polyhedron formed by the fingertips and the center of the palm.

Cylindrical precision. Equivalent to the previous case, but the hand frame is specified in the centroid of the polygon generated by the three fingertips.

One-finger frontal. The hand frame is set to the fingertip of one of the external fingers, whereas the other fingers remain slightly flexed.

One-finger precision. The same posture than the previous case, but the hand frame is placed on the inner part of the fingertip.

Table 4.1 Mapping between ideal preshapes, barrett hand preshapes and parallel jaw gripper preshapes

<table>
<tr><th>Ideal hand</th><th>Barrett Hand</th><th>Gripper</th></tr>
<tr><td>Hook power</td><td>Hook power</td><td rowspan="2">Pinch power</td></tr>
<tr><td>Cylindrical power</td><td>Cylindrical power</td></tr>
<tr><td>Hook precision</td><td>Hook precision</td><td rowspan="4">Pinch precision</td></tr>
<tr><td>Cylindrical precision</td><td rowspan="3">Cylindrical precision</td></tr>
<tr><td>Pinch</td></tr>
<tr><td>Lateral</td></tr>
<tr><td>One-finger frontal</td><td>One-finger frontal</td><td rowspan="2">One-finger</td></tr>
<tr><td>One-finger precision</td><td>One-finger precision</td></tr>
</table>

The mapping between the ideal preshapes and the Barrett hand preshapes is shown in Table 4.1. Each of the ideal hand preshapes has a correspondence with a Barrett hand preshape, with the exception of pinch and lateral. These postures require further dexterity and must be replaced by one of the feasible configurations. Concretely, in the case of the Barrett Hand, the pinch and lateral preshapes are mapped to the cylindrical precision posture.

4.2.2 Mapping to a Parallel Jaw Gripper

A parallel jaw gripper is the most simple form of robotic hand. It consists of two parallel jaws activated by a single motor on a prismatic joint. It offers very reduced dexterity. Thus, only three task-oriented hand preshapes are defined for this hand, as depicted in Figure 4.4 and described in the following:

Pinch power. The jaws are separated and the hand frame is set in the middle point of the line joining both fingers, at a near distance from the palm. This grasp is suitable for manipulating heavy objects, or parts where enough space is available in order to envelope the object.

Pinch precision. Equivalent to the pinch power case, but with the hand frame placed at the end of the jaws. This grasp is more suitable for small objects where precision is important.

One-finger. The gripper is closed and the hand frame is specified at the end of the jaws, with Z axis looking forwards. This preshape is useful for pushing actions.

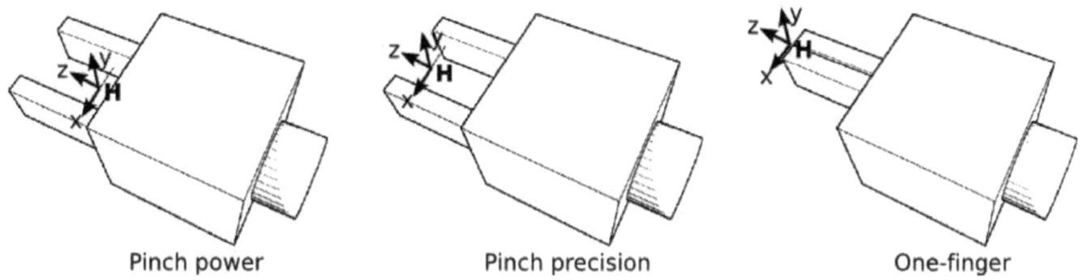

Fig. 4.4 The parallel jaw gripper task-oriented hand preshape

The correspondence between ideal preshapes and parallel jaw gripper preshapes is shown in Table 4.1. As the pinch grasp is almost the only possibility for this hand, all the configurations are mapped against the pinch preshape (either power or precision), with the exception of the ideal one-finger preshapes which can be also instantiated in the parallel jaw gripper.

4.3 Object Representation

The physical interaction planner needs a suitable task and object representation. From a description of the object and the task, the planner must be able to specify automatically all the elements of a physical interaction task. This section focuses on the object representation that we have adopted, whereas the next point will be devoted to the task description.

In our approach, the object is represented as a kinematic chain of shape primitives. Each object part is classified into an *object class* that indicates its function. The object class can be part of the object description, or can be assigned automatically according to the geometry and type of the joint.

4.3.1 Object Structural Model

The object structural model consists of geometric and kinematic information. It is composed of several shape primitives, linked in a tree hierarchy. Each node corresponds to an object part, approximated with a shape primitive. Each part is defined on its own reference frame, which is independent from the other parts. Each link represents the relative position between two object parts, including a motion constraint modelled by one of the following joints: prismatic, revolute and fixed. A prismatic joint allows one translational DOF between two object parts. A revolute joint involves a rotational DOF, whereas a fixed joint represents that the two object parts are rigidly linked. Therefore, an object is recursively defined as the union of several, possibly articulated, subobjects.

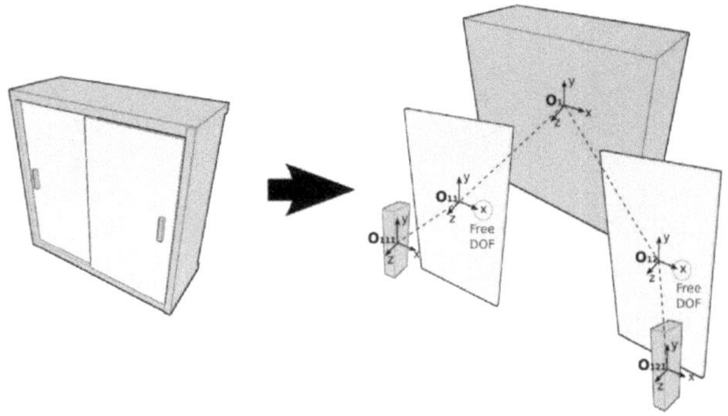

Fig. 4.5 Structural model of a cabinet

Only box shape primitives are considered because of the following reasons:

- A box is simple to describe geometrically, and can be used efficiently. The most common physical properties as the mass center, volume, dimensions, etc. can be computed easily.
- Any object geometry can be approximated by a bounding box. In addition, a box shape is specially well-suited for approximating furniture geometry.
- A box approximation is easier to obtain from sensor data than other geometries. See, for example, the approach of [75], where arbitrary object geometries are approximated by box primitives from a 3D point cloud obtained by vision or range sensors. The box-shape approximation is later used for planning grasps [74].

Figure 4.5 shows the structural model of a cabinet furniture. There is a base box corresponding to the cabinet structure, which contains two sliding doors as child objects. Each of the doors contain a handle which is rigidly linked to them. Each box is defined in its own reference frame. The pose of any part with respect to the parent is represented by an homogeneous transformation matrix. The motion constraints are indicated as free DOF's in the local frames, which must be chosen so that they are aligned with the *natural constraints*. For example, each of the doors can be moved freely along the X axis of their local frames.

4.3.2 Classification of Object Parts

The classification of object parts into categories has several advantages. First, the actions can be specified in an object-oriented way: each type of object allows a limited set of actions. All the possible tasks that can be performed on an object are encoded in the object itself. Second, it allows to establish the link between a

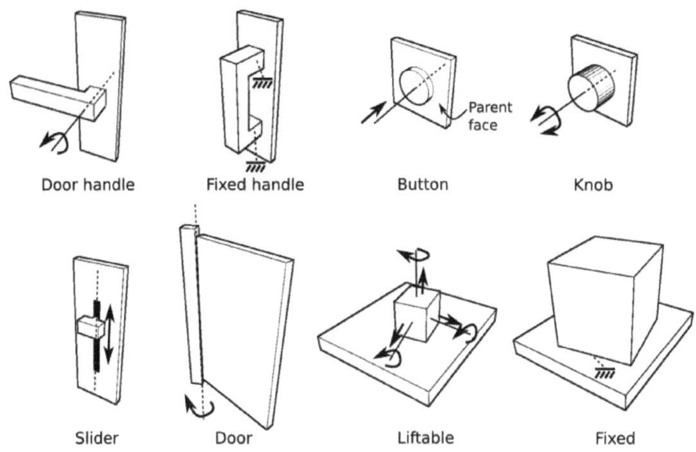

Fig. 4.6 The object classes considered by the physical interaction planner. Each object part can be automatically categorized into one class according to its geometry and joint type.

high-level task description and low-level actions. A high-level task description can be transformed into object actions, which are directly linked with the object degrees of freedom. And, third, the physical interaction planning problem can be divided into small subproblems, one for each object class.

We propose the categories depicted in Figure 4.6. Each object part can be classified into one of these categories according to the object geometry and the type of joint to which they are attached:

Door handle. An elongated box primitive attached to the parent object through a revolute joint.
Fixed handle. A hand-size box primitive rigidly attached to the parent object.
Button. A compact shape linked to the parent through a prismatic joint which axis is perpendicular to the parent face.
Knob. A compact shape linked to the parent through a revolute joint.
Slider. A compact shape linked to the parent through a prismatic joint which axis is parallel to the parent face.
Door. A box with two dimensions considerably larger that the other one, and attached to the parent object trough a revolute or prismatic joint with the joint axis along one of the largest dimensions.
Liftable. A small object which is not linked to a parent, and, thus, can be subject to free motion.
Fixed. A large object, not necessarily attached to a parent object, but for which motion is impossible due to its size, as, for example, the cabinet of figure 4.5.

The object category can be part of the object description or can be assigned by an automatic classifier. The former applies to an object recognition framework,

Table 4.2 Classification of object parts according to the motion constraint, physical properties and joint axis

Object class	Motion Constraint	Physical properties	Joint axis
Door handle	Revolute joint	Elongated, hand-size	Perpendicular to parent face
Fixed handle	Full-constrained	Elongated, hand-size	
Button	Prismatic joint	Compact, hand-size	Perpendicular to parent face
Knob	Revolute joint	Compact, hand-size	Perpendicular to parent face
Slider	Prismatic joint	Compact, hand-size	Parallel to parent face
Liftable	Free	Hand-size	
Door	Revolute joint	Planar	Parallel to parent face
Fixed	Full-constrained	Large	

where any object information can be stored in a database and accessed after the object is recognized. The latter can be used under an exploration framework, where the robot obtains a model by its own, without receiving external information.

The object classes considered by the planner can be distinguished in a confident way according to the object geometry and the type of joint to which they are attached. Table 4.2 shows a set of properties that can be used in order to classify an object part into one of the proposed categories.

4.3.3 Note on Object Models

The planner described in this chapter assumes there is an object structural model available, even though it can be described by simple box-shape primitives. The use of a common object representation allows to design a general planner valid for different geometries and tasks. Although a model is used by the proposed planner for building a physical interaction task specification, it is worth noting that other solutions could be possible, even without using object models.

We would also like to clarify that it is not necessary to have the complete object model in order to apply the physical interaction planner. For example, one could have a vision algorithm for recognizing door handles. From the output of this algorithm, it could be possible to build a simple box-shape model with the handle size, and classify it with the *Door handle* label. This simple model could be used as input to the physical interaction planner. The same idea is applied to other object types, such as knobs, buttons, etc.

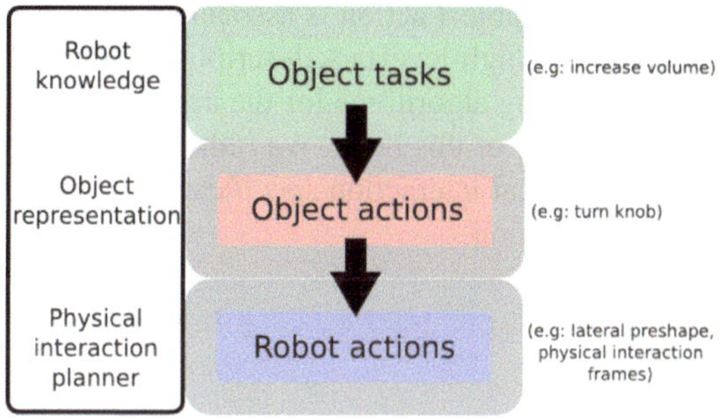

Fig. 4.7 Object tasks, object actions and robot actions

4.4 Task Description

Each class name is associated with a set of actions: the *object actions*. Each object action has a direct correspondence with a physical interaction task. Table 4.3 shows the actions defined for each object class, and its low-level description.

Table 4.3 The object actions associated to each object class and its description

Object class	Object action	Description
Door handle	Turn	Turns the handle in clockwise or counterclockwise direction
	Push	Push the handle forwards of backwards (pull)
Fixed handle	Push	Push along one of the following directions: forwards, backwards, right, left, up and down
Button	Push	Push the button forwards
Knob	Turn	Rotate the knob in clockwise or counterclockwise direction
Slider	Move	Translates the object part along the sliding mechanism
Liftable	Lift	A grasp on the object for transport purposes

Object actions provide an intuitive way to command physical interaction tasks at a middle-level language. However, commands can be given at a higher level. We define an *object task* as a name given to a sequence of one or more object actions. For example, "increase volume" is an object task involving the object action of "turn knob", "Switch on the television" would correspond to "push button", etc.

Thus, object tasks have a correspondence with object actions, and object actions correspond to low-level physical interaction tasks, as depicted in Figure 4.7. The link between object actions and physical interaction tasks is given by the physical interaction planner. The correspondence between object tasks and object actions must be stored in the robot by ad-hoc programming, or either learnt by demonstration, experimentation, etc.

Some object classes, like the *door* class, may not have object actions associated with them. Object parts categorized in such classes, can be used for describing high-level tasks, although the object action is associated with another part. For example, “open door” could be a high-level task description of the “pull handle” action.

The development of learning algorithms for the acquisition of knowledge about object tasks is out of the scope of this book. We rather focus on the transformation of object actions into a physical interaction specification, by means of the planner described in the following section.

4.5 Planning

There is a direct mapping between an object action and a physical interaction task. The physical interaction planner is in charge of this transformation. Taking as input an object representation and an action to perform on a given object part, the physical interaction planner defines the elements necessary for the specification of the physical interaction task, in terms of the framework described in the previous chapter.

4.5.1 *The Grasp*

According to expression 3.2, the grasp can be specified with a grasp preshape, $\mathscr{P}$, a hand frame, H, a grasp frame, G, the desired hand-grasp relationship, ${}^{H}\mathbf{M}_{G}^{*}$, and a matrix which selects the constrained DOF’s, $\mathbf{S_c}$.

The grasp preshape and the hand frame are jointly specified through the selection of a suitable ideal task-oriented hand preshape. The most suitable task-oriented hand preshape is chosen according to the object action and the task geometry. Table 4.4 shows the correspondence between an object action and the corresponding task-oriented hand preshape, depending on three geometric properties of the object part and the task: if there is enough space for enveloping the object with the fingers, whether the task direction is known or unknown, and whether the geometry of the box primitive is classified as elongated, compact, large or small.

The hand adaptors are then in charge of mapping the ideal preshape to a real posture on a specific hand. Next, the grasp frame is selected according to the following rules (see also Figure 4.8):

- In the case of prehensile grasps (cylindrical, pinch and lateral), its origin is placed on the center of the box primitive. In the case of non-prehensile grasps (hook and one-finger), its origin is set on the centroid of the box face which has its normal vector opposite to the task direction.
- The Z axis of the grasp frame must indicate the approaching direction. In the case of non-prehensile grasps, the approaching direction corresponds to the task direction.
- The rest of axes are chosen so that the desired state of the grasp link can be specified with the identity matrix, i.e. ${}^{H}\mathbf{M}_{G}^{*} = \mathbf{I}_{4\times4}$.

Table 4.4 Task-oriented preshapes assigned to each object action depending on the task parameters and object geometry

Object action	Gap	Direction	Size	Preshape
Door handle turn	Yes	Unknown		Cylindrical power
	Yes	Known		Hook power
	No	Unknown		Cylindrical precision
	No	Known		Hook precision
Fixed handle push	Yes	Unknown	Elongated	Cylindrical power
	Yes	Known	Elongated	Hook power
	No	Unknown	Elongated	Cylindrical precision
	No	Known (perpendicular)	Elongated	Cylindrical precision
	No	Known (parallel)	Elongated	Hook precision
	No		Compact	Lateral
Button push				One-finger frontal
Knob turn				Lateral
Slider move		Unknown		Lateral
		Known		One-finger precision
Liftable lift			Large	Cylindrical power
			Small	Cylindrical precision

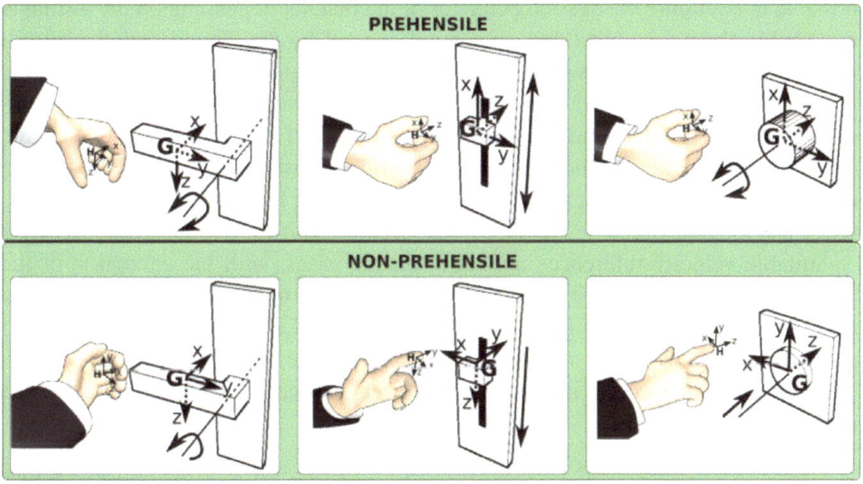

Fig. 4.8 Some examples of the grasp frame specification according to the task description and selected preshape. For prehensile preshapes, the grasp frame is placed on the center of the box primitive, with *Z* axis indicating the approaching direction. For non-prehensile preshapes, it is set to the box face which normal is opposite to the task direction.

The DOF's which are not constrained by the grasp depend on the selected preshape and grasp frame. In general, hook power, hook precision and cylindrical power preshapes do not constrain rotation around Y axis of the hand frame, whereas cylindrical precision, pinch and lateral preshapes could allow a rotation around X axis, and one-finger preshapes normally permit rotation around Z axis, and even around X and Y to some extent.

4.5.2 *The Task*

According to expression 3.3, the task specification is composed of the task frame, T, the velocity reference, $\mathbf{v}_T^*$, the force reference, $\mathbf{f}_T^*$, and the compliant selection matrix, $\mathbf{S_f}$.

The planner allows for two possibilities in the placement of the task frame. The first is to place it at the joint axis, as it would be required by the geometric compatibility property of the TFF. The other possibility is to set it at the same location than the grasp frame. The first case has the advantage that it is more coherent with the task geometry, and control would be simpler and more effective. However, an estimation of the task frame pose with respect to the grasp frame would be needed. The second case has the advantage of specification simplicity at the expense of control complexity. In addition, no knowledge about the TF pose would be required. It is possible to configure the desired behavior of the planner depending on the available knowledge about the object, but, in general, we opt for the second case, because it is more suitable for cases where geometric models are not available. For example, in some cases it could be desired to perform a door handle turning or pushing operation where only the handle has been recognized and no information is available about the door geometry and kinematics. Therefore, instead of relying on a motion model, we rely on robust sensor-based control in order to adapt the hand motion to the object kinematics.

Suitable velocity references are set in all the cases, with the exception of the *button push* action, where a force reference is set and the compliant selection matrix is defined accordingly. The exact velocity-force reference values should be also adapted to the particular case by means of experience-based learning algorithms. This represents one of the immediate extensions that could be implemented in the proposed planner.

4.6 Discussion and Conclusion

This chapter has described a planning algorithm that allows to automatically specify physical interaction tasks, taking as input an object description and the task to perform. Although the planner sets all the required elements that compose the definition of the grasp and the task in terms of the physical interaction framework, the automatic setting of suitable velocity-force references based on previous experience is still a pending problem.

The new concepts of *task-oriented hand preshapes*, *ideal hand* and *hand adaptors* have been introduced as a way to provide a general task-oriented planning algorithm that can be instantiated on different robotic hands, thus gaining in versatility.

Currently, the number of interaction tasks that the planner supports is limited by the set of object classes that have been considered. It would be possible to add new object classes and define object actions for them, or for the already existing ones. For example, the door class could allow a *knock* action. This action would require a new task-oriented hand preshape which is still not considered by the planner: a *fist preshape*. Most of the object parts found in our everyday environments can be classified into one of the proposed categories, but additional classes could be needed in other environments.

The planner makes use of object models built with box-shape primitives, although a complete model is not necessary in order to plan a physical interaction task. For planning an action on a handle, for example, it would be possible to recognize the handle with a vision system and approximate its shape with a bounding box, without the need to know its kinematics or the object to which it is attached.

The physical interaction planner proposed in this chapter contributes to the robot autonomy and versatility, as it allows to plan a variety of tasks in an automatic and hand-independent manner. On the next three chapters we will focus on the dependable execution of physical interaction tasks in terms of force, vision and tactile feedback, whereas chapter 8 will show specific examples of the application of this planner on different real objects and actions, as part of an assistive robot prototype.

Chapter 5
Physical Interaction: When Only Force Is Available

5.1 Brief Introduction

Force feedback represents a fundamental requirement for the success of any task that involves physical interaction with the environment. When combined with the touch sense, dexterous manipulation can be performed, even when deprived of visual information (indeed, blindness does not impede dexterity in humans). If tactile feedback is also removed, the use of force information allows humans to robustly perform manipulation tasks which do not require special dexterity, such as performing compliant motion. In the absence of force feedback, contact forces generated during the interaction with the environment cannot be measured. In this case, the success of a manipulation task depends on an accurately planned end-effector trajectory, generated by a pure motion controller based on internal models and the previous knowledge the robot has about the environment. Even if detailed models are available, it cannot be assumed that the environment is perfectly known. Therefore, we support that force feedback is a necessary requirement for the dependable execution of physical interaction tasks, and it is interesting to study how this information can be used for enabling reliable physical interaction when other sensors are not available.

When a robot is in contact with the environment, even a little positioning error can generate important interaction forces which can damage the robot or the objects. It is necessary to regulate those undesirable forces inside a safe range, by means of a suitable interaction control approach. Research on interaction control strategies started with *passive* and *active stiffness* control approaches [41, 158]. Passive stiffness control is performed by attaching to the robot end-effector a mechanically compliant device which is deformed according to the external forces. The main advantage of this approach is its control simplicity whereas the main limitation is that the stiffness properties cannot be easily modified: each compliant device is suitable for a particular task properties. In contrast, the active stiffness approach allows to control the desired stiffness at the end-effector according to the task requirements. *Impedance force control* [68] was introduced in order to take into account

M. Prats et al.: Robot Physical Interaction through the Combination of Vision, STAR 84, pp. 55–77.
springerlink.com

the full robot dynamics during position control when contact with the environment is expected. *Hybrid position-force control* [152] allows to specify at the same time explicit position and force references. A third control scheme, known as *external hybrid position-force control* [35, 141], integrates the advantages of the other two, and is characterized because of its implementation simplicity and better convergence properties. Although some variations to these schemes have been proposed in the literature, the original concepts are still valid in the force control community.

In this chapter, we address the problem of how to perform physical interaction tasks when only force information is available. In section 5.2 the main interaction control approaches existing in the literature are outlined, and its suitability for physical interaction tasks is studied in section 5.3. We also address the problem of how force feedback can be used in order to track the physical interaction frames, and, more concretely, the grasp link, which allows to transform the task specification into robot coordinates. Based on the framework for physical interaction, and the planner described in the previous sections, a real application where a humanoid robot is used in order to perform force-based physical interaction with different furniture in a kitchen environment is presented in section 5.4.

5.2 Interaction Control

In the literature we can distinguish between approaches that perform an indirect control of the interaction forces, those which control the contact force explicitly, and hybrid approaches [171]. The first group includes *passive* and *active compliance* control approaches. Among the active compliance schemes, it is worth mentioning *active stiffness control* and *impedance control*. These approaches control the interaction force indirectly, through adapted position control laws. In contrast, pure force control approaches allow to establish explicit force references as input to force control schemes build on existing motion controllers. *Hybrid position-force* approaches combine position control on some directions and explicit force control on the rest, being *hybrid parallel position-force* control and *external hybrid position-force* control the most important contributions.

In this section we outline the most common approaches in its position-force version. The equivalent velocity-force approaches can be obtained in most of the cases by replacing the position references and controllers with the corresponding velocity versions.

5.2.1 Impedance Control and Active Stiffness

Impedance control allows to specify a desired dynamic behavior of the robot when it is in contact with the environment [68]. Under constrained motion, the robot

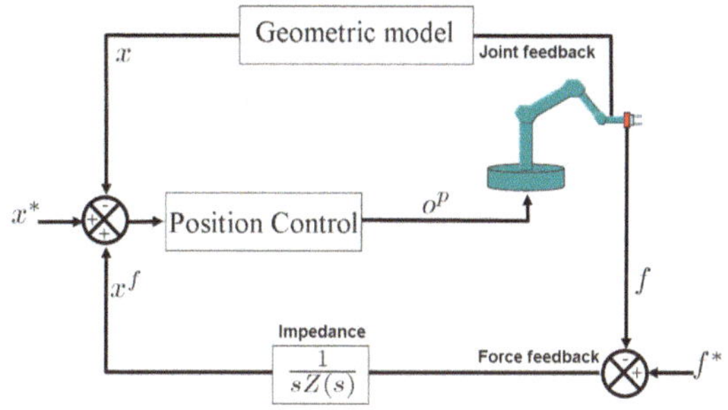

Fig. 5.1 General approach of the position-based impedance control. $\mathbf{o}_p$ is the output of the position controller.

end-effector position is related with the contact force through a programable mechanical impedance in the operational space, $\mathbf{Z}$, which can be represented in the frequency domain as:

$$\mathbf{F}(s) = s\mathbf{Z}(s)\mathbf{X}(s) \tag{5.1}$$

The robot is supposed to be equivalent to a mass-spring-damper second-order system whose transfer function is:

$$s\mathbf{Z}(s) = \Lambda s^2 + \mathbf{B}s + \mathbf{K} \tag{5.2}$$

where Λ, $\mathbf{B}$ and $\mathbf{K}$ are respectively the desired inertia, damping and stiffness matrices.

The classical impedance control approach requires a full characterization of the system dynamics. Unfortunately, the dynamic model of the manipulator is not always available, specially in industrial robots which usually implement off-the-shelf position-based controllers. In these situations, the desired impedance is normally used to modify a position control signal, as shown in Figure 5.1. When there is no contact force, i.e. $\mathbf{f} = 0$, the position trajectory is not modified, and the scheme performs pure motion control. Under a contact force, the position trajectory is modified according to the programmed mechanical impedance. This solution is valid as long as the low-level controllers are able to decouple the robot dynamics and mitigate the effects of external forces to some extent, which is the normal case in current robots working at slow velocities.

In addition, it is possible to consider only the stiffness matrix, $\mathbf{K}$, or the damping matrix, $\mathbf{B}$, leading to control schemes equivalent to the *active stiffness* [158] and *active damping* [189] approaches. In this book, we focus on active stiffness control, which establishes the following relationship between force and displacement:

$$d\mathbf{x} = \mathbf{K}^{-1}\mathbf{f} = \mathbf{C}\mathbf{f} \tag{5.3}$$

The inverse of the stiffness matrix is normally referred to as the *compliance matrix*, $\mathbf{C}$.

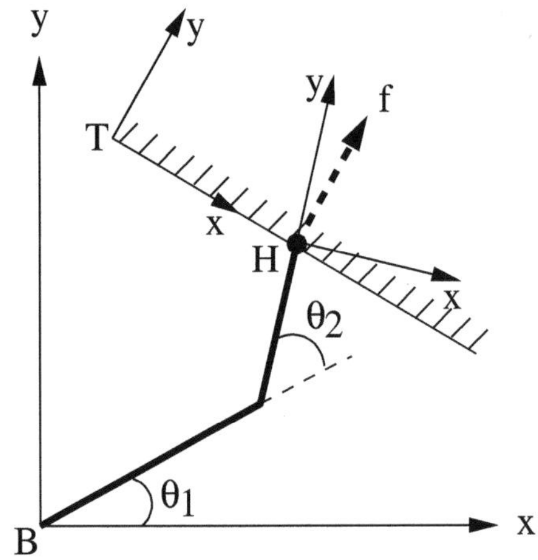

Fig. 5.2 A 2D surface following task with a 2 DOF manipulator. The TF is used as the compliance frame in the hybrid position-force approach. In this case, the *Y* direction is force-controlled while *X* axis is position-controlled.

The impedance-based approach performs an indirect control of the interaction force. Force feedback is used in order to update the dynamic behavior of the robot in the operational space, but it is not possible to establish a explicit force reference along a given direction. To overcome this problem, it is possible to introduce a force reference in the force feedback loop without affecting the control law stability, as it can be viewed as a reference trajectory modifier [123, 95]. This leads to the following expression, which will be used throughout this book:

$$d\mathbf{x} = \mathbf{K}^{-1}(\mathbf{f} - \mathbf{f}^*) \tag{5.4}$$

5.2.2 *Parallel Hybrid Position-Force Control*

The parallel hybrid position-force approach [152] has been normally used as the control scheme for tasks specified with the Task Frame Formalism. As it has been described in detail in section 3.2, the TFF divides the space into force-constrained and position-constrained directions. The hybrid position-force control assigns one different control mode to each subspace: position or velocity control to the force-constrained directions, and force control to the the position-constrained directions.

A *compliance frame*, which corresponds to the Task Frame, is associated to the object, so that its axes are aligned with the natural constraints. The compliance frame depends on the particular task to perform and, due to its alignment with the natural constraints, it decomposes the cartesian space into directions that must be force-controlled, and others which must be position or velocity-controlled. These directions are specified with the *compliant selection matrix*, $\mathbf{S_f}$ (see expression 3.3),

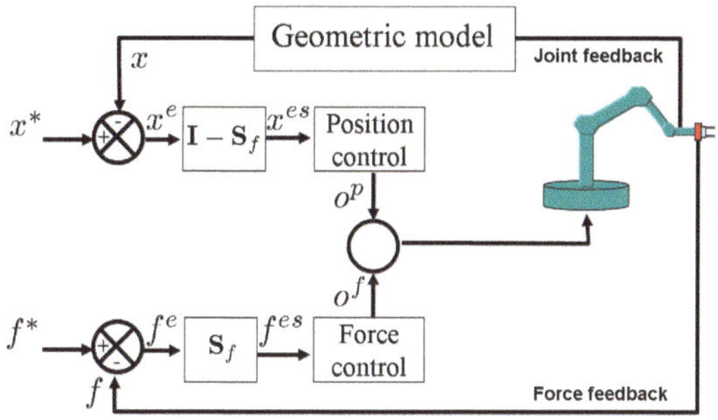

Fig. 5.3 General approach of the hybrid position-force control

which is a diagonal matrix where a diagonal value of 1 indicates a force-controlled direction, whereas a value of 0 indicates a position-controlled axis. Figure 5.2 shows an example of a surface following task. The compliance frame, T, has the Y axis perpendicular to the contact surface. Therefore, Y axis is a force-controlled direction, whereas X is position or velocity-controlled, i.e. $\mathbf{S_f} = \mathbf{diag}(0,1,0,0,0,0)$.

Figure 5.3 illustrates the general scheme of the parallel hybrid position-force control. One different controller is implemented for each subspace: a position controller for force-constrained directions, and a force controller for position-constrained directions. The control signals of both controllers are combined before sending them to the robot actuators. The position loop ensures a suitable trajectory of the robot in free space, whereas the force loop controls the motion in directions where contact with the environment is expected. More concretely:

- $\mathbf{x}^*$ and $\mathbf{f}^*$ are, respectively, the position and force references given in their respective frames.
- $\mathbf{q}$ and $\mathbf{f}$ are the current joint values the current force, after the corresponding calibration, filtering and gravity compensation.
- $\mathbf{x}$ is the cartesian position obtained from the direct geometric model of the robot.
- $\mathbf{x}^e$ and $\mathbf{f}^e$ are the position and force errors, defined as $\mathbf{x}^e = (\mathbf{x}^* - \mathbf{x})$ and $\mathbf{f}^e = (\mathbf{f}^* - \mathbf{f})$, respectively.
- $\mathbf{x}^{es}$ and $\mathbf{f}^{es}$ are the errors filtered through the compliant selection matrix, i.e. $\mathbf{x}^{es} = (\mathbf{I} \quad \mathbf{S_f})\,\mathbf{x}^e$, and $\mathbf{f}^{es} = \mathbf{S_f} \cdot \mathbf{f}^e$
- Finally, $\mathbf{o}_p$ and $\mathbf{o}_f$ are the outputs computed by the position and force control laws, and can be combined in different ways according to its nature (joint torques, operational space forces, etc.) [4].

The main advantage of this control approach is that it allows to implement the interaction tasks in a very intuitive manner, as it is based on the TFF. However, due to the reciprocity of the force-controlled and position-controlled subspaces, it is not possible to take into account forces into position-controlled directions, and vice versa. Therefore, the compliance frame must be known precisely, so that no disturbances

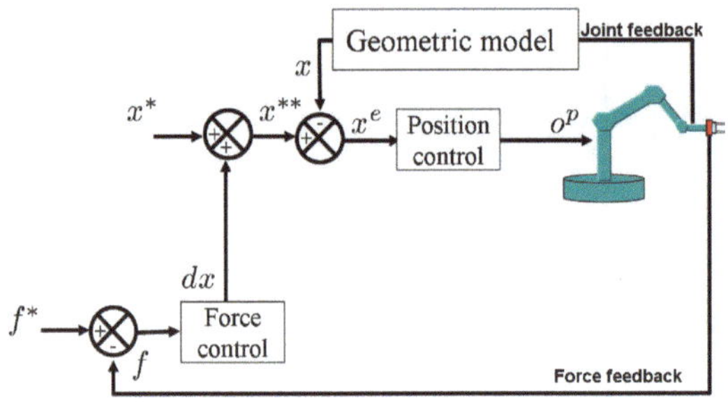

Fig. 5.4 General approach of the hybrid external position-force control

appear between the force-controlled and the position-controlled directions. This requires a detailed knowledge of the environment, which is normally not available in unstructured environments.

5.2.3 External Position-Force Control

In contrast to the hybrid parallel approach, the hybrid external position-force control [35, 141] allows simultaneous explicit force control and position control along the same direction. It consists of an inner position control loop enclosed inside an outer force controller. The output of the force controller modifies the reference value of the inner position controller, as shown in Figure 5.4.

A force error generates a control signal, $d\mathbf{x}$, that is added to the position reference, $\mathbf{x}^*$, in order to generate a new position reference, $\mathbf{x}^{**}$. The robot is finally controlled by the position controller, but the external force loop modifies the position controller input, thus adding a new control layer from which explicit force references can be set.

The main advantage of this approach is its implementation simplicity. In fact, in can be implemented on top of the existing original position controller of the robot. Furthermore, position and force control can be performed simultaneously on a same direction, thus overcoming the limitations of the hybrid parallel approach. Under some special circumstances, it has been shown that this approach is equivalent to the impedance control [89]. For a detailed comparison with other methods, please refer to [141].

5.3 The Role of Force Feedback during Physical Interaction

Among the contact information used by humans for grasping and interaction, the contact force and the pressure distribution on the hand or finger surface are of utmost importance. Some tactile sensors are able to provide both force and pressure information, but, normally, they cover only a small surface of the hand, such as the

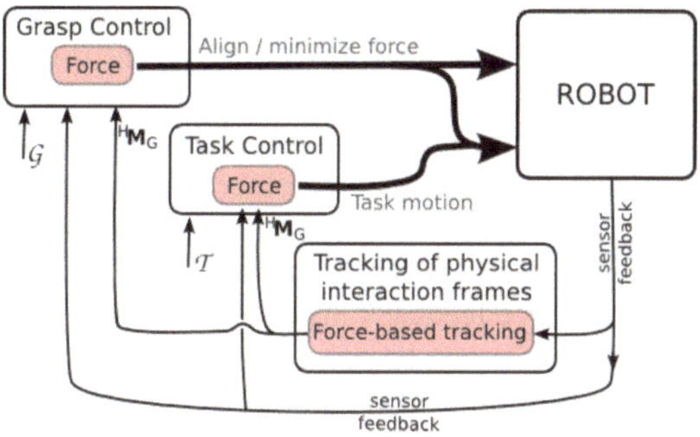

Fig. 5.5 The general force control scheme in terms of the physical interaction framework

fingertips or the palm. If the contact is performed with a part which is not covered by the tactile sensors, then there is no available information. In contrast, a force sensor placed at the robot wrist is able to provide the force generated by contacts on any part of the hand, ensuring that force feedback will be always available.

Force feedback is a necessary (and in some cases, sufficient) condition for enabling robust physical interaction. During grasping, force feedback can be used in order to detect contact with the object and adapt the hand motion accordingly. During interaction, a position-force control law can be adopted in order to perform compliant motion when the environment is not completely modeled or under geometric uncertainties.

In this work, vision and tactile sensors are considered as an additional help for improving force-based manipulation. Unlike force feedback, we cannot assume that robust visual and tactile information will be always present, the former because of the inherent difficulties associated to image processing in real life conditions, and the latter because of the very local nature of the information they provide. Therefore, it is necessary to address the problem of how to perform physical interaction robustly, when only force feedback is available.

Figure 5.5 shows the general physical interaction control scheme adapted to the case when only a force sensor can be used. Three aspects must be addressed: the force-based execution of the grasp, the force-based execution of the task, and the force-based tracking of the physical interaction frames.

5.3.1 Force Control for Grasping

Force feedback can be used during grasping mainly in two different ways. The first is as part of an event-based approach, where force conditions can represent transitions between different grasp states. The second use is as part of a closed control loop for alignment purposes.

When grasping an object from a table, for example, force information can be used in order to detect the contact of the fingers with the table surface. Similarly, when grasping a door handle, force feedback can provide information about the location of the door surface. In these cases, the contact force can be used as a condition for triggering the grasping action. Grasping failures can be also easily detected with a force sensor. For example, when moving an object to another place, a large discontinuity in the force signal can mean that the object has fallen during the pick and place action. In this case, the force discontinuity event feedback can be used as a condition for triggering an error recovering strategy.

Concerning the use of force feedback inside a control loop, one of the previous position-force control strategies can be adopted in order to adapt the hand position around the object. This is specially meaningful when grasping position-constrained objects, such as a door handle, for example. In this case, some fingers may make contact before the others, because of unexpected misalignments. These contacts generate forces which can be used in a position-force control approach for updating the hand position. The grasp controller must be active, not only during the grasping action, but also during the subsequent task. Force information can be used in order to keep a suitable grasp during motion, by aligning the hand with the grasped part via force minimization.

Special grasping strategies, combining position-force control approaches inside an event-based reasoning system can be implemented for grasping objects where basic control strategies are not appropriate. One example, that will be addressed in detail in the following section, is grasping a book from a bookshelf while it is standing among other books. Hybrid parallel position-force control will be combined with force-based events in order to perform the grasping action.

In general, force feedback provides a way to detect important events during the grasping action, such as the first contact, object missing, etc. and provides valuable feedback to position-force control approaches, which allow to perform grasping and compliant motion under considerable uncertainties in the hand-object pose. Thus, even in the absence of visual and tactile information, suitable force-based strategies can be defined for establishing a first grasp link, that will be later estimated and improved during the interaction phase.

5.3.1.1 An Example: The First UJI Librarian Robot

As an example of a special force-based grasping action, we describe in this section the book grasping strategy followed by the first version of the UJI Librarian Robot. Although this application will be described in detail in chapter 8, we show here the original book grasping strategy, using a parallel jaw gripper, endowed with special purpose fingertips, and attached to the robot end-effector via a force sensor.

The specification of the book grasping task is illustrated in Figure 5.6. In order to drive the hand frame towards the grasp frame, a special event-based grasp strategy has been implemented. Instead of inserting both jaws at the same time, which could be a difficult task, the left nail is inserted previously to the right one. For allowing this operation, the left jaw has been made longer than the right one. Some initial hypothesis must be assumed for the grasping to be feasible:

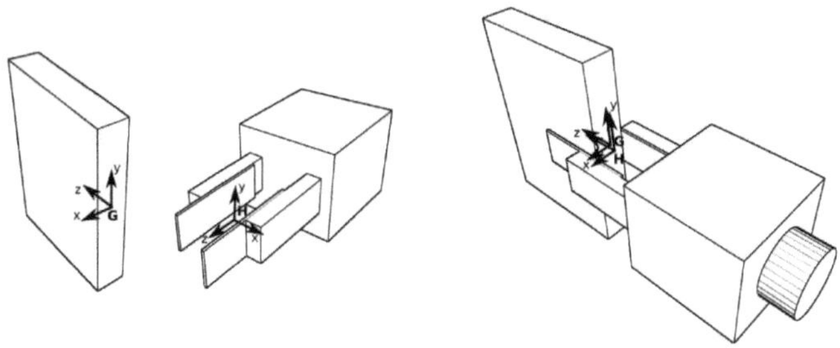

(a) The book is grasped from the front, with a pinch power preshape

(b) The grasp is completed when the hand frame reaches the grasp frame

Fig. 5.6 The book grasping task in terms of the physical interaction framework

- Books must be graspable by our robot arm. So, its physical properties (i.e. length, width, weight, etc.) will be always suitable to the available manipulation capabilities (i.e. geometry of the gripper, etc.).
- The spines of the books, oriented towards the outside the shelf, are all, approximately, in the same spatial plane.
- It is assumed that the books are not pressed together on the shelf in such a way as to impede the insertion of the gripper fingertips.
- We assume that the manipulator is adequately placed so that the camera plane is approximately parallel to the book plane.

The grasping process follows these steps:

1. Making contact with the spine of the desired book. A frontal contact force is used as the condition for triggering the next action (Fig 5.7a–b).
2. Looking for the left side of the book and inserting the left fingertip, by means of a hybrid parallel velocity-force control law, where $\mathbf{S_f} = \mathbf{diag}(0,0,1,0,0,0)$ (Fig 5.7c).
3. Opening the gripper and looking for the right side of the book while the left one remains in its previous location. This action is also performed with a velocity-force control law, with $\mathbf{S_f} = \mathbf{diag}(1,0,1,0,0,0)$ and $\mathbf{f}^*_x = 0$ (Fig 5.7d–e).
4. Gripping the book and taking it out of the bookshelf (Fig 5.7f–i).

The robot gripper starts moving towards the book in the direction perpendicular to the book plane, as shown in Figure 5.7a, until the force sensor detects frontal contact. At this point, the robot starts a surface following task, guided by a hybrid parallel position-force approach, in order to find the space at the left of the book and to insert the longer left nail into it. The velocity-force control law makes the robot apply a force forwards (force-controlled direction), while controlling the motion to the left (position-controlled direction). As long as the left finger lies on the book spine (Fig. 5.7b), the robot is applying a limited frontal force, while moving to the left.

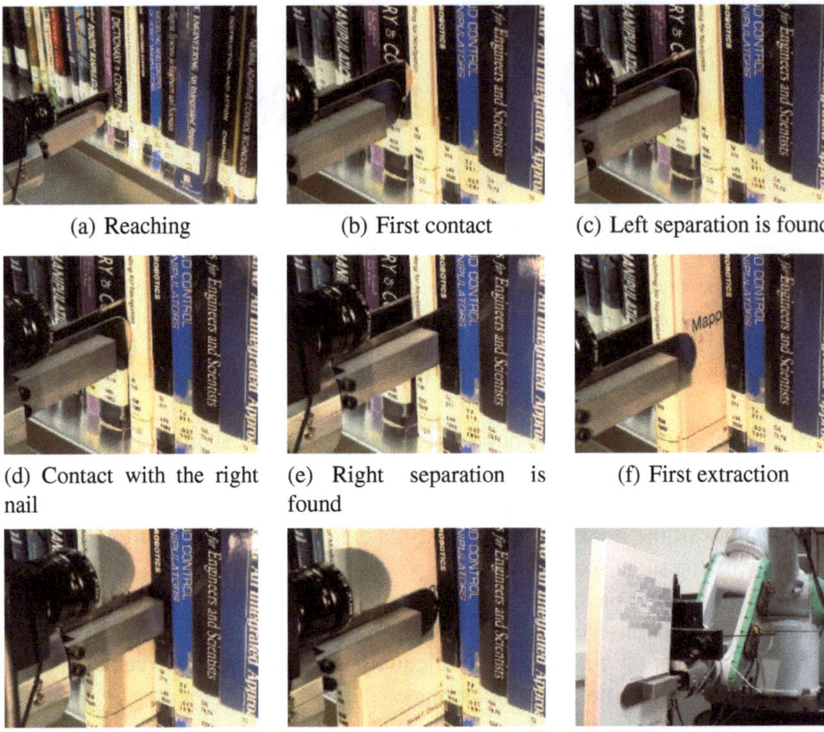

(a) Reaching (b) First contact (c) Left separation is found

(d) Contact with the right nail (e) Right separation is found (f) First extraction

(g) Regrasp (h) Second extraction (i) The book is taken out

Fig. 5.7 Book grasping with a parallel jaw gripper

When the left separation is found (Fig. 5.7c), the robot does not sense any frontal force, and the position-force control law makes it move forwards until another force is detected due to contact with the right nail. When the robot detects that its left fingertip has advanced 2 cm with respect to the position where initial contact was made, it is considered that the first nail has been inserted.

For inserting the right fingertip, the robot still tries to move the fingers forwards. If the right separation is not found, it will sense an opposite force (Fig. 5.7d), but this time, instead of moving to the right, it opens the gripper some millimeters. As one nail is already inserted, when opening the gripper the robot senses some lateral forces that it tries to minimize using the same velocity-force controller, where both the lateral and frontal directions are force-controlled. When the separation at the right is found, the gripper can advance freely, and it stops opening. Similarly to the left case, when the robot detects that the gripper has advanced 4 cm with respect to the initial position where first contact was made, it assumes that both jaws have been successfully inserted (Fig. 5.7e). Simple extraction movements and re-grasping complete the grasping part (Figs. 5.7f–i).

5.3.2 Force Control for Interaction

During interaction, force feedback is absolutely necessary for performing compliant motion in a safe and robust manner. Under compliant motion, the robot end-effector trajectory is constrained by the task geometry. It cannot be assumed that a complete and precise knowledge about the constrained trajectory is available. Even if it is available, small errors in the relative robot-object positioning can generate considerable interaction forces that can damage the robot or the environment. Interaction control allows to detect these forces and update the hand position in order to regulate them inside a suitable range.

As in the case of grasping, the analysis of the force signal during interaction can be useful for detecting interesting task states and switching between actions accordingly. For example, an excessive force during a door opening task is probably indicating that the door is locked. Similarly, if a considerable force appears after turning a handle some degrees, it could be indicating that the limit of the handle mechanism has been reached, and that we should switch to the pulling action.

Force feedback during interaction can also be helpful in order to estimate the grasp link, or computing an approximation of the hand-grasp transformation under compliant motion. Indeed, the force generated by the environment on the robot hand as a reaction to the robot motion, depends on the particular motion constraints of the manipulated object. By exploring the task directions which do not generate reaction forces, it is possible to estimate the particular object mechanism, and, therefore, the hand-object transformation, as it is explained in the following section.

5.3.3 Force-Based Estimation of the Grasp Link

In section 3.4.3.3, we have reported the importance of building a sensor-based estimation of the grasp link, or the relative transformation between the hand and the grasp frame. The grasp link allows to transform the task specification into robot coordinates, which is necessary for performing the interaction task.

Although force feedback provides very local information about the contact, it seems apparently sufficient for performing the most common actions, specially those where the grasp constrains all the relative hand-object motion. Humans are able to turn a door handle and opening the door even when deprived of more complete information, such as that provided by the vision sense. By means of force feedback, we are able to adapt our hand motion to the particular constrained trajectory, without the need of vision, and even without knowledge of the object.

The same approach can be adopted in a robotic environment. For illustrating it, we consider a robot which has already performed a prehensile grasp (possibly underconstrained) on a handle, and we assume that the robot has an initial estimation of the task direction, tangent to the particular constrained trajectory, as shown in Figure 5.8. We can design, for example, a hybrid external position-force control scheme, with a force reference along the task direction and zero force reference on the rest of axis. This makes the robot push along the estimated task direction while regulating

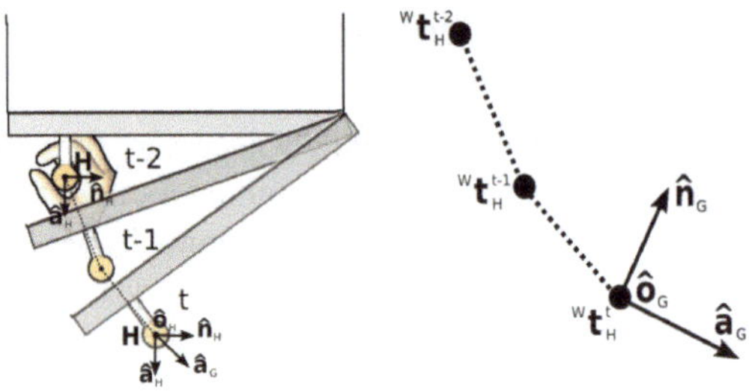

Fig. 5.8 The grasp frame can be estimated by monitoring the hand trajectory when performing the task under force control

the lateral forces to zero. Thus, the robot hand starts pushing along one direction and updates its position in the tangent directions according to the forces generated by the constrained trajectory of the particular mechanism.

During this compliant motion, the hand frame position can be monitored in order to estimate the constrained trajectory, and, ultimately, the grasp frame pose with respect to the hand frame. For this, the hand trajectory is sampled at a given time interval, and the last three points are used in order to estimate a vector tangent to the constrained trajectory, as shown in Figure 5.8.

More concretely, let $^W\mathbf{t}_H^t$, $^W\mathbf{t}_H^{t-1}$ and $^W\mathbf{t}_H^{t-2}$ be the last three points sampled on the trajectory followed by the hand frame under the particular force control approach, given on a fixed frame, W. A second-order approximation of the tangent to the trajectory at point $^W\mathbf{t}_H^t$ can be computed as:

$$\mathbf{a}_G = \frac{^W\mathbf{t}_H^{t-2} - 4\,^W\mathbf{t}_H^{t-1} + 3\,^W\mathbf{t}_H^t}{2} \tag{5.5}$$

$$\hat{\mathbf{a}_G} = \frac{\mathbf{a}_G}{||\mathbf{a}_G||} \tag{5.6}$$

It can be assumed that this vector is aligned with one of the axis of G. Therefore, it can be used in order to build an estimation of the grasp frame pose with respect to the grasp frame. For this, we need two additional unit vectors, orthogonal to $\hat{\mathbf{a}_G}$. The first is $\hat{\mathbf{n}_G}$, built as:

$$\hat{\mathbf{n}_G} = \frac{\hat{\mathbf{o}_H} \times \hat{\mathbf{a}_G}}{||\hat{\mathbf{o}_H} \times \hat{\mathbf{a}_G}||} \tag{5.7}$$

The second is $\hat{\mathbf{o}_G}$, built as:

$$\hat{\mathbf{o}_G} = \hat{\mathbf{a}_G} \times \hat{\mathbf{n}_G} \tag{5.8}$$

The grasp frame pose with respect to W is then built as:

$$^W\mathbf{M}_G = \begin{pmatrix} \hat{\mathbf{n}_G} & \hat{\mathbf{o}_G} & \hat{\mathbf{a}_G} & ^W\mathbf{t}_H^t \\ 0 & 0 & 0 & 1 \end{pmatrix} \tag{5.9}$$

and an estimation of the grasp link is then obtained as $^H\mathbf{M}_G = \left(^W\mathbf{M}_H\right)^{-1} \cdot {}^W\mathbf{M}_G$.

In conclusion, we have shown how to estimate the grasp link during an interaction task, when force is the only available sensor. The main concept is to push along the initial estimation of the task direction at the same time that the hand naturally adapts to the particular constrained trajectory by means of a suitable interaction control law. The trajectory followed by the hand under force control is monitored in order to estimate the grasp frame pose, which is then used to build a new grasp link estimation. This strategy is put into practice in the force-based application described in the following section.

5.4 Practical Implementation: The Armar-III Humanoid Robot

In this section, we present a humanoid robot able to perform compliant physical interaction tasks with furniture commonly found in household environments, using only force feedback. The general framework for task description and sensor-based execution, developed previously in chapters 3 and 4, is adopted for this purpose, providing versatility to the robot, which is able to adapt to several different cases, without being specifically programmed for a particular task. Robustness to uncertainties during task execution is guaranteed by a force-torque sensor placed in the robot's wrist, which is in charge of estimating the grasp link, thus adapting the robot motion to the particular task, without using previous knowledge about the object mechanism. A total of 8 DOF's are controlled, making the task execution highly redundant, thus allowing the use of auxiliary secondary tasks. Grasp redundancy is also exploited in order to further increase the space of secondary motion, which allows to keep a comfortable arm posture while performing the main task. The purpose is to show a first real application of the physical interaction framework under force-based control strategies. This work was developed during a short stay of 5 months in Karlsruhe University (Germany).

5.4.1 Experimental Environment

The Armar-III humanoid robot [6], built up by the Collaborative Research Center 588 in Karlsruhe, has been used for this application. This robot is composed of a humanoid torso mounted on a holonomic mobile platform, and has a total of 43 DOF's. A complete kitchen has been built in order to validate the robot capabilities in a realistic household environment (see Figure 5.9). The kitchen contains all of the elements that can be found in any common kitchen, like a dishwasher, a fridge, a microwave, different cupboards, drawers, etc. For this particular application, we focus on force-guided reliable physical interaction of the Armar-III humanoid robot with articulated furniture found in the kitchen, such as opening doors and drawers.

Fig. 5.9 The Armar-III humanoid robot

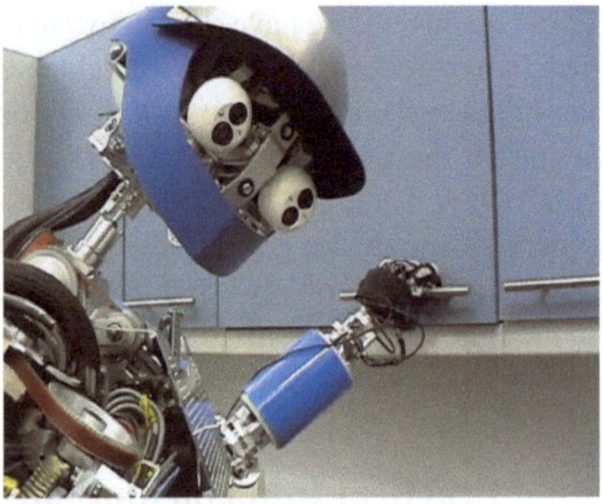

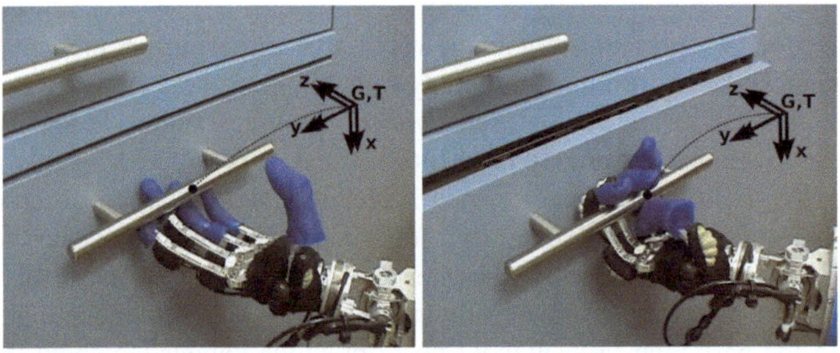

Fig. 5.10 The grasp and task frames are set to the handle. The task is specified as a velocity reference along negative *Z* axis. A cylindrical power grasp is performed.

5.4.2 Physical Interaction Planning and Specification

For the particular constrained object that the robot has to manipulate, we assume that an initial estimation of the handle pose is available. This estimation can be obtained from the robot localization algorithms, or through a previous vision-based recognition and pose estimation step. In addition, it is also assumed that a simple box-shaped model of the handle is available. As all the handles in the kitchen have the same shape, a single model is valid for all of them. This model is classified with the *Fixed handle* label, and is given as input to the physical interaction planner, together with the *Push backwards* object action. This simple task description is valid for the different doors and drawers found in the kitchen.

Following the planning rules explained in section 4.5, the hook power task-oriented preshape is selected, which is mapped to a cylindrical power preshape by the hand adaptor implemented for the Armar-III hand, as shown in Figure 5.10. The task and grasp frames are set to the estimated position of the handle, and the task velocity is set to a negative value along *Z* axis, $\mathbf{v}_T^* = (0,0,-20,0,0,0)^T$ mm/s, which was chosen experimentally. As the force reference is not required for these kind of tasks, the planner sets $\mathbf{S_f} = \mathbf{0}$. Grasp redundancy is allowed in the rotational DOF's, i.e. $\mathbf{S_c} = \mathbf{diag}(1,1,1,0,0,0)$. Normally, only a rotation around the handle axis should be allowed as a grasp redundant DOF. However, as the robot's hand is highly compliant, and all the DOF's are force-controlled at the low-level, the three cartesian rotational DOF's are set to be controlled by the secondary task (avoiding joint limits), and not by the main task. This allows to increase the range of the task without reaching joint limits.

Due to grasp redundancy, the task frame is not necessarily rigidly attached to the hand frame. The task frame, according to its definition, must be always aligned with the natural decomposition of the task. Thus, sensors must provide an estimation of the task frame position and orientation during the task execution. In this application, we make use of the force-based estimation of the grasp link described in the previous point in order to estimate the task frame (as G and T are placed at the same position, ${}^H\mathbf{M}_G = {}^H\mathbf{M}_T$).

It is worth mentioning that this task specification avoids the use of the particular mechanism model. The same task description is used for the different doors and drawers, independently of the particular size or hinge position. The force-based estimation of the grasp link allows to automatically adapt the hand trajectory to the particular constrained motion.

5.4.3 The Physical Interaction Controller

According to the general control guidelines outlined by the physical interaction planner in section 3.4.3, a suitable physical interaction controlled must be composed of a task controller, in charge of performing the task motion, and a grasp controller, devoted to perform the grasping action, and to keep a suitable grasp link during interaction. Both controllers must receive continuous information about the state of the physical interaction frames, obtained from sensor feedback.

In this section, we present a physical interaction controller for the Armar-III humanoid robot that fulfills these requirements. A velocity-force controller, taking into account joint and grasp redundancy, is used for both the grasping and the interaction steps. Redundant degrees of freedom are managed by a secondary task, based on a cost function minimization approach. This allows the robot to robustly interact with different furniture while adopting a comfortable joint and grasp link configuration.

As force sensor readings are affected by the gravity force, we first present an approach to subtract it, based on an estimation the hand mass and gravity center. Next, a basic cartesian velocity controller is presented, which is then extended with joint and grasp redundancy management, relying on the force-based tracking of the grasp link state. Finally, force feedback is integrated in the control law, following an active stiffness approach.

5.4.3.1 Force Sensor Calibration

When using a force sensor in the wrist, the raw force readings $\mathbf{f}_r$ are affected by the gravity force, $\mathbf{g}$, acting on the hand for a particular end-effector position. In order to get the real task forces, the gravity force must be subtracted from the readings. The gravity wrench, $\mathbf{f}_g$, acting on a body with mass m and center of mass $\mathbf{p} = (c_x, c_y, c_z)^T$ can be expressed as:

$$\mathbf{f}_g = \begin{pmatrix} m\mathbf{g} \\ -m\mathbf{g} \times \mathbf{p} \end{pmatrix} = \begin{pmatrix} m\mathbf{g} \\ -[m\mathbf{g}]_\times \mathbf{p} \end{pmatrix} = \begin{pmatrix} \mathbf{D}(\mathbf{g}) & \mathbf{0}_3 \\ \mathbf{0}_3 & -[\mathbf{g}]_\times \end{pmatrix} \cdot \mathbf{c}$$

where $\mathbf{D}(\mathbf{g})$ is a 3×3 diagonal matrix whose diagonal is $\mathbf{g}$, i.e. $\mathbf{D}(\mathbf{g}) = \mathbf{diag}(\mathbf{g})$, $[\mathbf{g}]_\times$ is the skew-symmetric matrix defined by the components of $\mathbf{g}$, and $\mathbf{c} = m(1,1,1,c_x,c_y,c_z)^T$. In order to subtract the gravity force, the vector $\mathbf{c}$ must be estimated. This can be done from a set of n raw force readings, $\mathbf{f}_r^0 \ldots \mathbf{f}_r^n$, at known end-effector positions by solving the following system through least-squares minimization:

$$\begin{pmatrix} \mathbf{D}(\mathbf{g^0}) & \mathbf{0}_3 \\ \mathbf{0}_3 & -[\mathbf{g^0}]_\times \\ \vdots & \vdots \\ \mathbf{D}(\mathbf{g^n}) & \mathbf{0}_3 \\ \mathbf{0}_3 & -[\mathbf{g^n}]_\times \end{pmatrix} \cdot \mathbf{c} = \begin{pmatrix} \mathbf{f}_r^0 \\ \vdots \\ \mathbf{f}_r^n \end{pmatrix}$$

where $\mathbf{g}^i$ is the gravity vector expressed in the force sensor frame at the ith end-effector position ($i \in [0 \ldots n]$). Once $\mathbf{c}$ has been estimated, the real force $\mathbf{f}$, independent of the gravity force, can be computed in the following way:

$$\mathbf{f} = \mathbf{f}_r - \begin{pmatrix} \mathbf{D}(\mathbf{g}) & \mathbf{0}_3 \\ \mathbf{0}_3 & -[\mathbf{g}]_\times \end{pmatrix} \cdot \mathbf{c}$$

5.4.3.2 Cartesian Velocity Control

Tasks are naturally defined in the 3D cartesian space, where objects lie [73]. To open a drawer, for example, one must pull along a straight line in cartesian space. For opening a door, an arc in cartesian space must be followed. For this reason, the framework is based on the computation an end-effector cartesian velocity, that must be later transformed into joint space by a suitable cartesian controller.

It is well-known that joint velocity and cartesian velocity are related by:

$$\mathbf{v}_F = \mathbf{J}_F \dot{\mathbf{q}} \tag{5.10}$$

where $\mathbf{J}_F$ is the jacobian matrix for the particular robotic system, given in a frame F, $\dot{\mathbf{q}}$ is a vector with joint velocities, and $\mathbf{v}_F$ is the end-effector cartesian velocity, given in the frame F. A particularly important case is when F is set to the end-effector frame, i.e. $F = E$. In this case, the jacobian is given in the end-effector frame, and, thus, $\mathbf{v}_F = \mathbf{v}_E$ represents the cartesian velocity at the end-effector corresponding to the joint velocity $\dot{\mathbf{q}}$, given at the current end-effector frame.

If the jacobian matrix is given at the end-effector frame and can be inverted, the robot can be controlled at a desired end-effector cartesian velocity with:

$$\dot{\mathbf{q}} = \mathbf{J}_E^{+}\mathbf{v}_E \tag{5.11}$$

where $\mathbf{J}_E^{+}$ represents the pseudo-inverse of the Jacobian matrix.

5.4.3.3 Joint Redundancy

A robot manipulator is redundant at the joint level when the number of available joints is greater than those needed to reach a given position and orientation in the cartesian space (a 6D pose). In the context of this application, 8 DOF are used in the Armar-III humanoid robot: 7 in the arm, and 1 in the hip (yaw). Therefore, two redundant degrees of freedom exist at the joint level, allowing the robot to reach a given cartesian pose with many different joint configurations.

We adopt the well-known *gradient projection method* (GPM) for joint redundancy management [170]. The general approach is to project a secondary task into the null-space of the first task, as following:

$$\dot{\mathbf{q}} = \mathbf{J}_E^{+}\mathbf{v}_E + (\mathbf{I} - \mathbf{J}_E^{+}\mathbf{J}_E)\mathbf{e_j} \tag{5.12}$$

where $\mathbf{J}_E$ is the arm jacobian at the end-effector, $(\mathbf{I} - \mathbf{J}_E^{+}\mathbf{J}_E)$ is the null-space projector, and $\mathbf{e_j}$ is a secondary task cost vector, which is normally computed as the gradient of a cost function $\mathbf{h}(\mathbf{q})$, i.e. $\mathbf{e_j} = \frac{\partial \mathbf{h}}{\partial \mathbf{q}}$. There are many different possibilities for the cost function: increase manipulability, minimize energy consumption, etc. In our case, we compute $\mathbf{h}$ in order to avoid joint limits, thus keeping a natural arm posture. Concretely, $\mathbf{e_j}$ is taken as:

$$\mathbf{e_j}_i = \frac{\partial \mathbf{h}}{\partial \mathbf{q}_i} = \begin{cases} -\frac{q_i - q_i^u}{q_i^r} & \text{if} \quad q_i > q_i^u \\ -\frac{q_i - q_i^l}{q_i^r} & \text{if} \quad q_i < q_i^l \\ 0 & \text{in other case} \end{cases}$$

where q_i^u, q_i^l and q_i^r are, respectively, the upper bound, lower bound and range for joint i. By upper and lower bound, we understand the joint value at which the joint limit avoidance secondary task is activated (not the mechanical limit).

5.4.3.4 Grasp Redundancy

The number of redundant DOF's can be further increased if grasp redundancy is also considered. Grasp redundancy is defined in the operational space, as the number of

DOF's which do not need to be controlled for a particular grasp. For example, when grasping a handle, a rotation around the handle axis can be considered as grasp-redundant if the grasp and handle geometry allow it, as it was shown in the example of Figure 3.4.

For task execution, velocity and force must be controlled at the task frame. Thus, the task jacobian must be computed from the end-effector jacobian, as:

$$\mathbf{J}_T = {}^E\mathbf{W}_T^{-1} \cdot \mathbf{J}_E \tag{5.13}$$

where ${}^E\mathbf{W}_T$ is computed from the estimated transformations between the physical interaction frames, specially the grasp link, i.e. ${}^E\mathbf{W}_T = {}^E\mathbf{W}_H \cdot {}^H\mathbf{W}_G \cdot {}^G\mathbf{W}_T$ (see also expression 3.6). As the task and grasp frames are set to the same position, a modified task jacobian can be computed taking into account grasp redundancy, as:

$$\mathbf{J}_T^r = \mathbf{S_c} \cdot \mathbf{J}_T \tag{5.14}$$

being $\mathbf{S_c}$ the diagonal selection matrix introduced by the physical interaction framework in section 3.4.2.1, which selects the degrees of freedom in the operational space constrained by the grasp. Then, equation 5.12 can be expressed in task coordinates as:

$$\dot{\mathbf{q}} = \mathbf{J}_T^{r\,+} \mathbf{v}_T^* + (\mathbf{I} - \mathbf{J}_T^{r\,+} \mathbf{J}_T^r)\mathbf{e_j} \tag{5.15}$$

With this new expression, $\mathbf{e_j}$ is projected not only on the joint-redundant DOF's, but also on the grasp-redundant ones. Therefore, more DOF's are available for the secondary task, allowing the robot to perform the main task, while effectively performing auxiliary motion.

5.4.3.5 The Advantages of Grasp Redundancy

Joint and grasp redundancy allow to increase the range of motion of the robot for a given task. Figure 5.11 shows a simulated Armar-III humanoid robot opening the door of the dishwasher. Each of the images show the maximum opening reached under five different conditions: 7 DOF's without joint redundancy management, 7 DOF's with joint redundancy, 8 DOF's without joint redundancy, 8 DOF's with joint redundancy, and 8 DOF's with joint and grasp redundancy.

The best results are obtained in the last case, 8 DOF's with joint and grasp redundancy, where the maximum opening angle is reached, as shown in Figure 5.12a. When using 7 DOF's, joint limits are reached very quickly, specially if no secondary task is used at the joint level. The opening angle increases as more degrees of freedom are used and joint redundancy is exploited. Figure 5.12b shows the evolution of the cost vector norm during the opening task with the different strategies. Under 7 DOF's without secondary task, there is a high cost during the whole task. Results are better if joint redundancy is exploited, although cost is still high and the limit of the workspace is reached quickly. The use of 8 DOF's increases the workspace, and

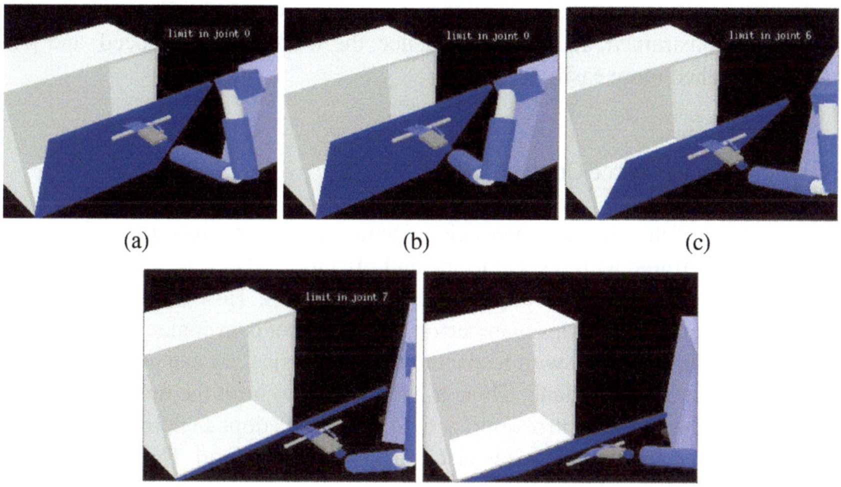

(a) (b) (c) (d) (e)

Fig. 5.11 Final state of the task and arm configuration for each of the different approaches: (a) 7 DOF's without joint redundancy management, (b) 7 DOF's with joint redundancy, (c) 8 DOF's without joint redundancy, (d) 8 DOF's with joint redundancy, and (e) 8 DOF's with joint and grasp redundancy

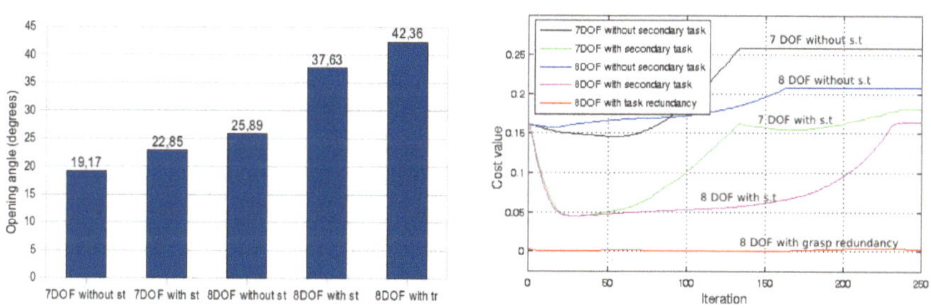

(a) Opening angle (b) Norm of the cost vector

Fig. 5.12 Comparison of the performance according to the number of DOF's, joint redundancy and grasp redundancy

results are much better in this case, specially when introducing grasp redundancy management, where the cost value is close to zero during the whole operation.

Grasp redundancy allows to release the constraint on some degrees of freedom between the hand frame and the grasp frame. As it is shown in Figure 5.11, when grasp redundancy is exploited, the relative orientation between the hand and the handle is not fixed (Fig. 5.11e). It can be controlled by the secondary task in order to perform auxiliary motion (joint limit avoidance in this case). In contrast, when task redundancy is not considered, the relative orientation between the hand and the handle remains fixed during the interaction task. The six cartesian DOF's at the end-effector are constrained, and, as consequence, the workspace is reduced, and joint limits are reached more easily.

5.4.3.6 Velocity-Force Control Law

If the tracking of the physical interaction frames, and more concretely, the estimation of the task frame with respect to the end-effector, ${}^{E}\mathbf{M}_T$, was perfect, the robot could perform the task without any kind of force feedback. However, in practice, the task frame estimation contains many errors. As the robot is in contact with the environment, any misalignment with respect to the exact trajectory can generate considerable forces on the robot hand. Thus, it is necessary to limit the interaction forces through a suitable force control approach. For this, we adopt an active stiffness approach where the force reference is also introduced in the loop. The velocity-force control law running in the robot for compliant cartesian velocity control with joint and grasp redundancy is the following:

$$\dot{\mathbf{q}} = \mathbf{J}_T^{r\,+}\mathbf{v}_T^* + (\mathbf{I} - \mathbf{J}_T^{r\,+}\mathbf{J}_T^r)\mathbf{e_j} + \mathbf{J}_T^+\mathbf{K}^{-1}({}^{T}\mathbf{D}_F \cdot \mathbf{f}_F - \mathbf{f}_T^*) \tag{5.16}$$

being $\mathbf{K}$ the stiffness matrix, and ${}^{T}\mathbf{D}_F$ the wrench transformation matrix between frames F and T (i.e. ${}^{T}\mathbf{D}_F = {}^{T}\mathbf{W}_F^T$) [89]. The matrix ${}^{T}\mathbf{M}_F$ can be computed as ${}^{T}\mathbf{M}_F = {}^{E}\mathbf{M}_T^{-1} \cdot {}^{E}\mathbf{M}_F$, where ${}^{E}\mathbf{M}_F$ relates the force sensor frame with the end-effector. Finally, $\mathbf{f}_F$ is the measured force at the force sensor frame, and $\mathbf{f}_T^*$ the desired force at the task frame.

The first two terms allow to perform the task motion and the auxiliary task based on the coarse estimation of the task frame, which may introduce some unexpected forces. The last term adds a correction to the joint velocities depending on desired and current forces, allowing to perform the task compliantly, even when important errors exist in the task frame estimation, ${}^{E}\mathbf{M}_T$ (and, thus, in $\mathbf{J}_T^+$). As the robot is designed to work in human environments, task velocities have been set to small values, so that dynamics issues can be neglected.

The control equation 5.16 is used for both the grasp and the task controller. The grasp is performed by setting $\mathbf{v}_T^* = \mathbf{f}_T^* = \mathbf{0}$. When the hand envelopes the handle, contacts appear and the arm configuration is automatically adapted in order to minimize external forces, making the hand adapt naturally to the handle. After that, $\mathbf{v}_T^*$ is set to the task reference.

5.4.4 Experimental Results

Figure 5.13 shows the Armar-III humanoid robot interacting with four different elements found in a common kitchen: a fridge, a drawer, a dishwasher and a cupboard. Each of them have different size and even different mechanism. In the case of the dishwasher, the hinges are located at the bottom part, unlike the fridge and cupboard

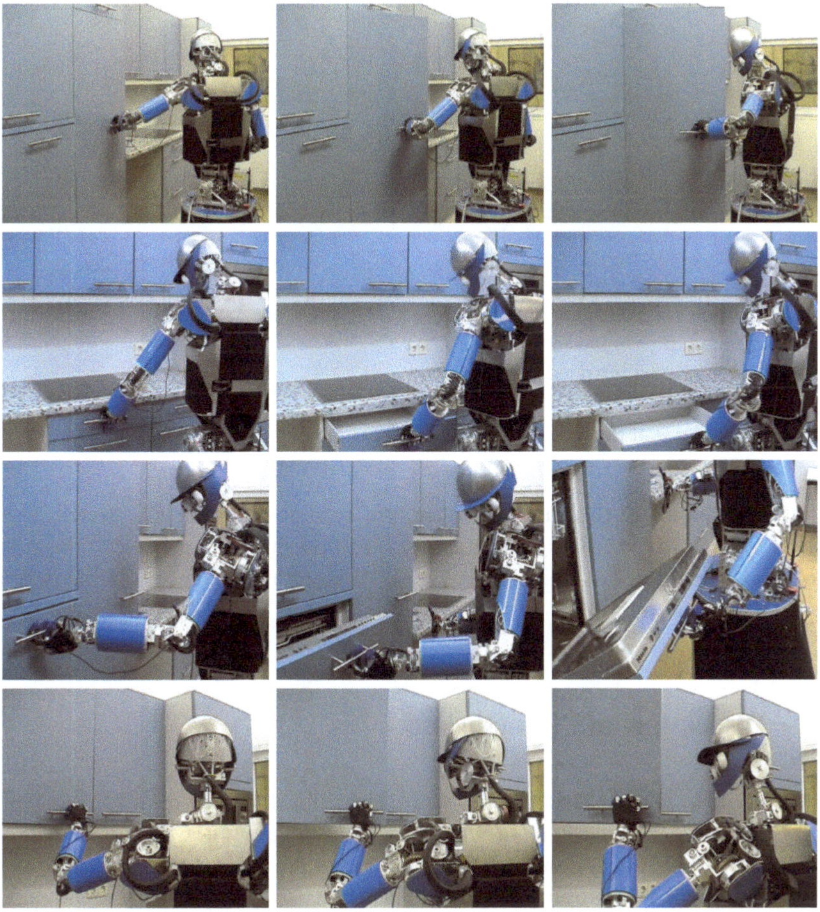

Fig. 5.13 The Armar-III robot interacting with the kitchen furniture: fridge, drawers, dishwasher and cupboards

where the hinges are at the left. In the case of the drawer, the mechanism is not a revolute joint, but a prismatic one. However, the task description is the same for all of them: pulling the handle. Thanks to the on-line force-based estimation of the hand-to-handle relative pose, the robot is able to adapt its motion to the particular case, without being specifically programmed for any particular task.

In addition, because of the high number of redundant DOF's and the grasp and joint redundancy management, the humanoid robot is able to perform the tasks in a very natural way, by adopting a comfortable arm posture, far from the joint limits, if possible. Figure 5.14 shows the values of the secondary task cost vector when opening the door of the dishwasher. When the task starts, cost is decreased to a value close to zero due to internal motion, using the redundant DOF's. The cost remains small during task execution, until the end, when it starts increasing. This

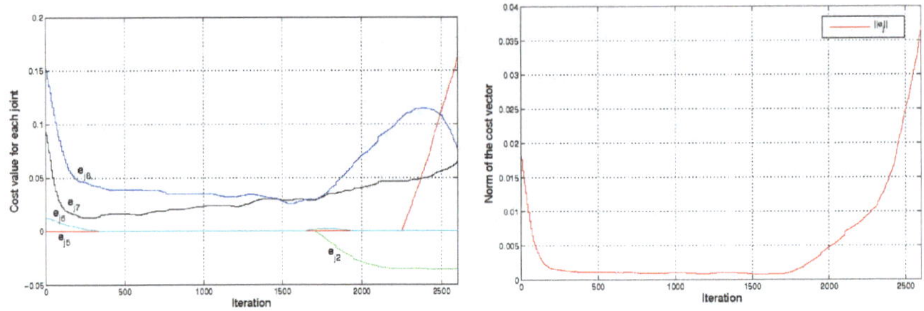

(a) Cost value for each joint. Joint 1 corresponds to the hip yaw and the 8 to the wrist pitch. The joints not illustrated have zero cost.

(b) Norm of the cost vector.

Fig. 5.14 Influence of the secondary task when opening the door of the dishwasher

point corresponds to the case when the arm is almost completely stretched (see Figure 5.13, third row, third column, and joint 5 cost in Figure 5.14a). At this point, the task is considered as finished, as very little internal motion is possible in order to avoid the joint limits.

5.5 Discussion and Conclusions

In this chapter, we have discussed how force feedback can be used for the execution of physical interaction tasks when other sensors are not available. In fact, we cannot assume that vision information will always provide correct and accurate information. Similarly, current commercial tactile sensors only cover a small part of the hand. Contacts produced outside the covered area cannot be detected. On the contrary, a force sensor placed on the robot wrist can robustly detect any contact force generated on the hand, even in situations when the other sensors fail. Therefore, it is important to study how force feedback can be used for performing robust manipulation, independently of the rest of sensors.

We have outlined the main existing methods for position-force control, namely impedance control, hybrid parallel control and external hybrid control. We have also analyzed how these methods can be used, in terms of the physical interaction framework, for performing the grasp and the constrained task. Concretely, force feedback can be used during grasping for detecting contact with the object and adapt the hand motion accordingly. Regarding interaction, existing position-force control approaches can be adopted in order to perform the constrained trajectory under geometric uncertainties. Interestingly, force feedback can also be combined with position information in order to track the physical interaction frames, allowing the robot to naturally adapt to the particular constrained trajectory without planning the specific motion.

These concepts have been illustrated with two examples: grasping a book with a parallel jaw gripper, and interacting with kitchen furniture with an advanced humanoid system. In the first one, a book is grasped from a bookshelf while standing among other books, and using exclusively force information. In the second application, a humanoid robot interacts with the fridge, dishwasher, cabinets and drawers of a kitchen, without relying on its geometric models. For this, the force-based tracking of the physical interaction frames is applied, jointly with a position-force control strategy that exploits joint and grasp redundancy for increasing the robot workspace.

In the next chapters, we will show how vision and tactile feedback can be added on top of force-based control strategies with the purpose to provide further reliability and versatility. Vision will be used for directly tracking the physical interaction frames and assist force-based manipulation accordingly. Tactile information will be used, when available, for accurately controlling a subset of cartesian directions when vision and force feedback do not provide enough information. However, in the absence of robust visual and tactile feedback, the methods presented in this chapter provide a robust solution for basic constrained manipulation.

Chapter 6
Physical Interaction: Adding Vision

6.1 Brief Introduction

Whereas the previous chapter has been devoted to the study of how physical interaction tasks can be executed using exclusively force feedback, we consider now the advantages of adding vision sensors. Computer vision can provide a powerful way of sensing the environment and can potentially reduce or avoid the need of structuring it. The purpose of this chapter is to study the additional value that vision feedback can provide to force-based physical interaction.

Vision and force sensors provide complementary and very different information. Vision sensors can provide a rich knowledge about the position and orientation of objects in the environment, whereas force sensors can provide local information to eventually correct the 3D trajectory of the robot end-effector when it performs constrained motion. Each one is useful for a different phase of the physical interaction task: vision allows accurate part alignment in partially unknown and/or dynamic environments, before and after contact is performed, whereas force enables compliant motion during contact.

Several researchers have studied vision-force control techniques with the purpose of getting the benefits of both sensors. This chapter firstly presents a review of existing vision-based control methods, concretely in section 6.2, and pays special attention to object pose estimation strategies, necessary for estimating the position and orientation of the manipulated objects in the environment. In this line, we present a new *active pose* estimation method, suitable for large objects which do not fit in the robot's field of view. This method is specially interesting in the service robotics context, where the robot frequently needs to manipulate some parts of large and fixed objects like doors, cupboards, etc. Next, a review of vision-force control approaches is presented in section 6.3, and a new control method is proposed: the *external hybrid vision-force control*. We then discuss in section 6.4 how the previously outlined vision-force control techniques can be applied in the basis of the physical interaction framework, for both aspects of a physical interaction task.

M. Prats et al.: Robot Physical Interaction through the Combination of Vision, STAR 84, pp. 79–110.
springerlink.com

Finally, in section 6.5, we show an practical implementation of the theoretical aspects in a real robot, performing vision-force physical interaction in the real world.

Results show how the use of vision feedback allows to retrieve the pose of the objects in the environment, thus avoiding the need to work in structured scenarios. In addition, this pose can be used to directly estimate the grasp link state, thus detecting any misalignment between the hand and the manipulated object, and making the robot adapt to the particular constrained trajectory through a suitable vision-force control strategy.

6.2 Visual Control of Manipulators

Motivated by the desire to reduce or avoid the need of structuring the environment, the use of visual observations to control robot motion has been extensively studied in the last years [23, 24, 109]. Typical applications of vision-based methods include the positioning of a robot with respect to an object, known as *visual servoing*, using either an external or end-effector-mounted camera.

Several visual servoing methods have been proposed in the literature [76]. On the one hand, they can be classified into *dynamic look-and-move* versus *direct visual servoing* approaches. Dynamic look-and-move systems consider that a low-level joint controller exists, accepting position or velocity inputs in the cartesian space. In contrast, direct visual servoing completely replaces the robot controller by a vision-based control law which directly generates control signals at the low-level. All the vision-based experiments considered in this monograph are classified into the first group. On the other hand, the visual servoing systems can also be classified, according to the nature of visual features they use, into *position-based*, *image-based* and *hybrid* approaches. The first one is based on the computation of a 3D cartesian pose (position and orientation), and normally requires a 3D model of the object and a calibrated camera [113]. In contrast, image-based visual servoing does not need a full model of the target, because the control loop is directly closed in the image space [50], providing increased robustness to modeling errors and noise perturbations with respect to position-based approaches [22]. More recently, several researchers have explored hybrid approaches which combine euclidean and image information in order to design globally stabilizing control [30, 110]. Finally, a third classification exists, depending on whether visual observations of the end-effector are introduced in the control law, or, in contrast, only the target object is observed. The first approach, known as *endpoint closed-loop*, allows to work with relative measurements, and, thus, it is robust to hand-eye calibration errors. On the contrary, the second approach, denoted as *endpoint open-loop*, requires an accurate knowledge of the camera position with respect to a robot-related frame, although the tracking of the end-effector is avoided.

6.2.1 Visual Servoing Fundamentals

All the visual servoing methods are based on a common concept: a set of visual features must reach a desired state by means of suitable robot motion. Consider the vector $\mathbf{s} = (\mathbf{s}_1, \mathbf{s}_2, \ldots, \mathbf{s}_n)^T$, containing the visual observations used as input to the control scheme, where $\mathbf{s}_i$ is a m-dimensional vector containing a particular subset of the visual features. If the 3D features corresponding to the visual observations are motionless, the time derivative of $\mathbf{s}_i$ is:

$$\dot{\mathbf{s}}_i = \frac{\partial \mathbf{s}_i}{\partial \mathbf{c}_O} \frac{d\mathbf{c}_O}{dt} = \mathbf{L}_i \mathbf{v}_C \tag{6.1}$$

where $\mathbf{c}_O$ is the camera position with respect to the visual target and $\mathbf{v}_C$ is a 6-dimensional vector denoting the camera velocity. The $m \times 6$ matrix, $\mathbf{L}_i$, is the interaction matrix related to $\mathbf{s}_i$, which links the variation of the visual observations to the camera velocity. If we consider the time derivative of the full feature vector, $\mathbf{s}$, the corresponding interaction matrix is $\mathbf{L} = (\mathbf{L}_i, \cdots, \mathbf{L}_n)^T$ and:

$$\dot{\mathbf{s}} = \mathbf{L} \cdot \mathbf{v}_C \tag{6.2}$$

In order to control the robot motion so that the visual features reach the desired state, an error function, $\mathbf{e}$, is normally defined as:

$$\mathbf{e} = \mathbf{s} - \mathbf{s}^* \tag{6.3}$$

where $\mathbf{s}^*$ is the desired value of the visual features vector, $\mathbf{s}$. Forcing an exponential decrease of the error in the form of $\dot{\mathbf{e}} = -\lambda \mathbf{e}$, leads to the following control law:

$$\mathbf{v}_C = -\lambda \widehat{\mathbf{L}}^+ (\mathbf{s} - \mathbf{s}^*) \tag{6.4}$$

where $\widehat{\mathbf{L}}^+$ represents the pseudo-inverse of a chosen model of the interaction matrix. The expression of $\mathbf{L}$ for different sort of image features can be found in [50]. It is well know that such system is locally asymptotically stable in a neighborhood of $\mathbf{s}^*$ if and only if $\mathbf{L}\widehat{\mathbf{L}}^+$ is a positive defined matrix [22].

In order to compute the control law of equation 6.4, it is necessary to provide an approximated interaction matrix, $\widehat{\mathbf{L}}$, which depends on the particular visual features chosen for the application. Image-based approaches take 2D features directly extracted from the images, whereas position-based methods use 3D features obtained through reconstruction methods.

It is known that humans use 3D information rather than 2D features for vision-based manipulation [73]. In addition, the use of 3D information allows to easily integrate the vision part with the physical interaction framework and planning algorithms, which are naturally defined in the 3D cartesian space. For these reasons, most of the visual servoing methods adopted in this work follow a position-based approach, which in most cases require obtaining the 3D position and orientation of the objects in the environment. In the following section, the most common pose estimation methods are outlined, including one important contribution of this book: an *active pose estimation* approach, which performs suitable camera movements in order to obtain the full 3D pose of large objects which do not fit completely in the camera view.

6.2.2 Vision-Based Object Pose Estimation

In a manipulation context, vision is normally used in order to obtain an estimation of the object position and orientation that allows the robot to perform a specific action with it [92, 161, 176, 45]. Pose estimation techniques in robot vision can be classified in *appearance-based* vs. *model-based approaches* [101]. Appearance-based methods work by comparing the current view of the object with those stored in a database containing previously acquired views from multiple angles. The main advantage of these methods is that they do not need a 3D object model, although a previous tedious process must be performed in order to include a new object in the database. In contrast, model-based methods need a 3D model as input, but better accuracy and robustness is obtained. In addition, the use of model information allows for anticipating events like object self-occlusions. Some approaches consider a combination of both methods, like [92], where an appearance-based method is used first for getting an initial pose estimation, which is then used as initialization for a model-based algorithm. For the vision-based approaches of this work, we assume that an object structural model, as it has been described in section 4.3, has been previously acquired through an object recognition method. Therefore, we focus on model-based pose estimation techniques, paying special attention to those which can also deal with articulated structures.

There are two main methods in the literature for model-based pose estimation and tracking of articulated objects, both based on full-scale non-linear optimization. The first, introduced in [44], is formulated from the Lie algebra point of view, whereas the second, proposed in [28, 27], is based on the Virtual Visual Servoing (VVS) method [111]. Both approaches implement robust estimation techniques and have shown to be very suitable for real-time tracking of common articulated objects in real environments. A comparison between both methods is reported in [26], where it is shown that both formulations are equivalent, although some differences in performance can appear at run time.

Throughout this book, we focus on the VVS approach [27, 111], mainly for its computational efficiency and because it is based on a solid background theory, i.e. 2D visual servoing, which convergence conditions, stability, robustness, etc. have been widely studied in the visual servoing community [76]. In addition, almost any kind of visual feature can be used and combined with this approach (points, lines, ellipses, etc.), as long as the corresponding interaction matrix can be computed. Different examples of the interaction matrix for the most common features can be found in [50]. In the following sections, a more detailed view of the VVS-based pose estimation approach is presented. Its limitations are reported, and a new active pose estimation approach is proposed for dealing with these limitations.

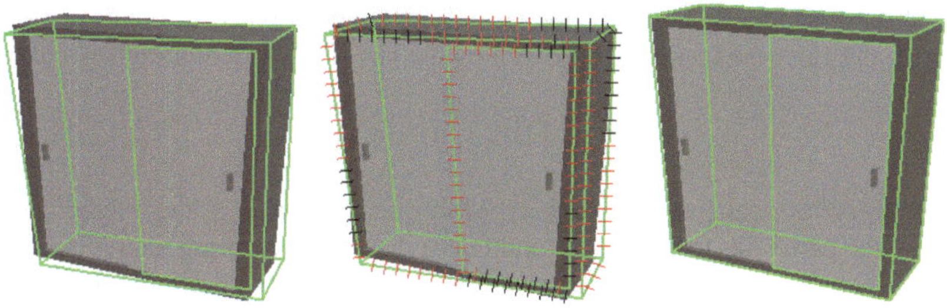

(a) Initial estimation (b) Point-line distance minimization (c) Final estimation

Fig. 6.1 An outline of the VVS-based pose estimation approach based on the point-to-line distance feature. A feature vector is built from the distances between the projected edges and high-gradient points searched along the edge normals, at the sampling interval. The goal of the non-linear minimization is to reduce all the distances to zero.

6.2.2.1 Virtual Visual Servoing

The concept of the VVS approach, developed in [111], is to apply visual servoing techniques to a virtual camera, so that a set of object features projected in the virtual image from a model, match with those extracted from the real image. Under this approach, the pose estimation and tracking problem can be seen as equivalent to the problem of 2D visual servoing [28], which has been extensively studied in the visual servoing community [76].

Taking as input a 3D features model, and an initial estimation of the camera pose in object coordinates, denoted as a pose vector, $\mathbf{c}_O$, the idea is to project a set of 3D features of the object model into a virtual image of the object, taken from the virtual camera position, $\mathbf{c}_O$. This virtual image is compared with the real one, and a vector of visual features is generated, denoted by $\mathbf{s}(\mathbf{c}_O)$.

In the particular vision-based applications described in this monograph, we make use of two different kinds of visual features: the point-to-point distance using special markers (practical implementation of this chapter), and the point-to-line distance on the real object edges (practical implementation of chapters 7 and 8). In the point-to-point distance, points are tracked in the image, and the 2D euclidean distances between the estimated positions and the tracked ones are used as visual features. In the point-to-line distance, the edges of the object model, projected as lines in the virtual image, are sampled at regular intervals, and a search for a strong gradient is performed in the real image, in a direction perpendicular to the projected line, as shown in Figure 6.1. For each match, the point-to-line distance feature [26] is computed and stored in the feature vector. The desired feature vector is given by $\mathbf{s}^* = 0$, which represents the case when all the edges of the object model are projected on strong gradients, and, ideally, the virtual camera position corresponds to the real one. In any case, the control law governing the virtual camera motion is given by:

$$\mathbf{v}_C = -\lambda \left(\widehat{\mathbf{D}}\widehat{\mathbf{L}_s}\right)^+ \widehat{\mathbf{D}}(\mathbf{s}(\mathbf{c}_O) - \mathbf{s}^*) \tag{6.5}$$

where $\mathbf{v}_C$ is the virtual camera velocity, λ is a control gain, $\widehat{\mathbf{L}_s}$ is the interaction matrix for the particular feature set, $\mathbf{s}$, and $\widehat{\mathbf{D}}$ is a diagonal weighting matrix computed by iteratively re-weighted least squares, which is a robust estimator for dealing with outliers.

6.2.2.2 Virtual Visual Servoing on Articulated Objects

In [27] an approach for pose estimation and tracking of articulated objects, based on the VVS method and the kinematic set concept, was presented. In this approach, the articulated pose is estimated directly from the visual observation of the object parts, leading to an efficient method that eliminates the propagation of errors through the kinematic chain. The only condition is that joint parameters must be decoupled in the minimization of the objective function. This can be accomplished by performing the minimization in object joint coordinates instead of in the camera space. Let $\mathbf{s}_1(\mathbf{c}_O^1)$ and $\mathbf{s}_2(\mathbf{c}_O^2)$ represent the perceived visual features on both parts of an articulated object composed of two links and one joint, and $\mathbf{s}_1^*$ and $\mathbf{s}_2^*$ be the desired values for those features, with $\widehat{\mathbf{L}_{s1}}$ and $\widehat{\mathbf{L}_{s2}}$ representing the corresponding interaction matrices. Then, the articular pose can be estimated by applying the following image-based control law:

$$\begin{pmatrix} \mathbf{v}_1 \\ \mathbf{v}_2 \end{pmatrix} = -\lambda \widehat{\mathbf{A}} \left(\widehat{\mathbf{D}}\widehat{\mathbf{H}}\right)^+ \widehat{\mathbf{D}} \begin{pmatrix} \mathbf{s}_1(\mathbf{c}_O^1) - \mathbf{s}_1^* \\ \mathbf{s}_2(\mathbf{c}_O^2) - \mathbf{s}_2^* \end{pmatrix} \tag{6.6}$$

$$\widehat{\mathbf{H}} = \begin{pmatrix} \widehat{\mathbf{L}_{s1}} & \mathbf{0} \\ \mathbf{0} & \widehat{\mathbf{L}_{s2}} \end{pmatrix} \widehat{\mathbf{A}}$$

$$\widehat{\mathbf{A}} = \begin{pmatrix} \widehat{{}^C\mathbf{W}_O}\mathbf{S} & \widehat{{}^C\mathbf{W}_O}\mathbf{S}^\perp & \mathbf{0} \\ \widehat{{}^C\mathbf{W}_O}\mathbf{S} & \mathbf{0} & \widehat{{}^C\mathbf{W}_O}\mathbf{S}^\perp \end{pmatrix}$$

where $\widehat{{}^C\mathbf{W}_O}$ represents the twist transformation matrix from the camera frame to the object joint frame, and $\mathbf{S}^\perp$ is a constraint matrix which depends on the type of joint [27]. Finally, the virtual camera velocities, one for each link, are given by $\mathbf{v}_1$ and $\mathbf{v}_2$. It is worth mentioning that the object joint frame, O, and the constrained matrix, $\mathbf{S}^\perp$, can be obtained from the object structural model used by the physical interaction planner (see section 4.3), which makes possible the integration of this approach with the physical interaction planner.

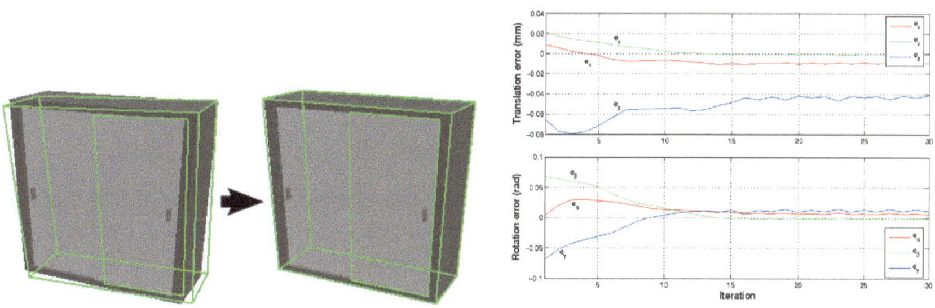

(a) The cabinet is seen from a far position with a coarse initial estimation that is corrected

(b) The pose estimation converges to the real value up to a small error

Fig. 6.2 VVS convergence in the case where there is enough information in the image and the interaction matrix is full rank

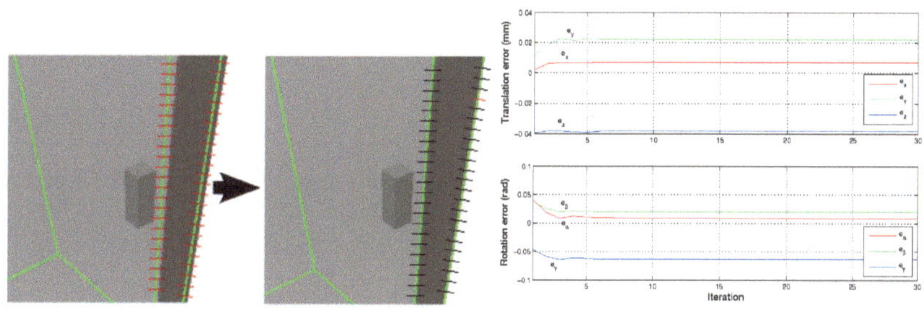

(a) The virtual edges converge to real ones

(b) The pose estimation does not converge

Fig. 6.3 VVS convergence in the case where there is not enough information in the image and the interaction matrix is not full rank

6.2.2.3 Limitations

The main limiting factor affecting the convergence of the VVS method is local minima, which depends on the kind of features considered and its observability in the image. In general, the interaction matrix must be full rank in order to be able to compute a solution for equations 6.5 and 6.6.

This is the typical case for objects that are fully observable from the camera point of view. Figure 6.2 shows a simulation in which a cabinet is seen from a far position. In order to study the convergence of the VVS method in this case, important errors have been manually introduced in the initial camera position. Figure 6.2b shows how VVS is able to improve the initial estimation, and converges to the real pose, up to a small error. The reason is that a rich set of visual features is available, which ensures the full rank of matrix $\widehat{\mathbf{L}_s}$.

However, this is difficult to achieve in a manipulation environment, where the robot is close to the target object, and only a small part of it is visible. As an example,

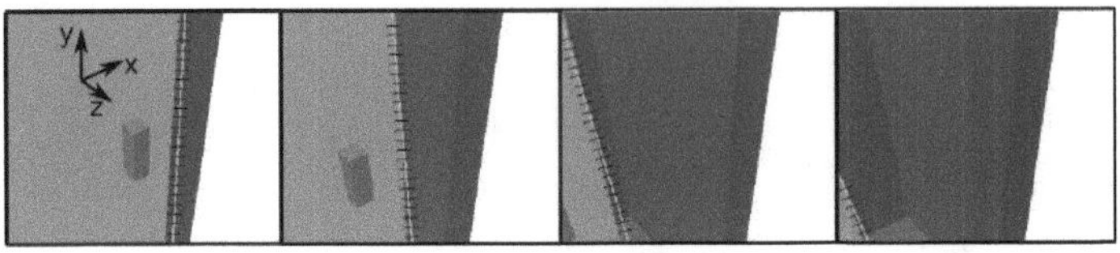

Fig. 6.4 Tracking of the articulated part when the parent object is considered fixed at a position with a manually introduced error. As the prismatic joint only involves 1 DOF, tracking can be performed even with a rank-deficient interaction matrix.

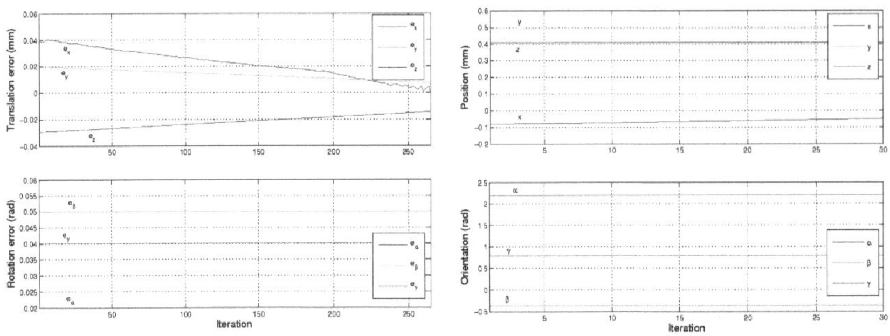

(a) The pose error

(b) Evolution of the camera pose in object coordinates

Fig. 6.5 Tracking of a sliding door along one DOF

Figure 6.3 shows the same cabinet seen from a position suitable for manipulation purposes. In this case, only the right edges of the object are visible. Although the handle features could also be considered in the simulation environment, we have not used them because of the difficulty to extract them robustly in a real case, apart from the fact that, during manipulation, the handle features are normally occluded by the hand. In addition, the handle can be different from one door to another, whereas the door edges are always present. In this experiment, the interaction matrix, computed from the set of point-to-line features extracted from the visible edges, is not full-rank, which means that there is ambiguity and multiple solutions are possible. It is worth noting that this is not a limitation of the method, but a result of the particular visual conditions. In fact, without considering the handle, it would be ambiguous also for the human eye. Figure 6.3b shows how the VVS method converges to a situation where some DOF's are even worse than the initial estimation, even though the projected edges correspond to the real ones, as depicted in Figure 6.3a. Therefore, local minima can appear if the feature set is not rich enough.

Fortunately, this problem can be detected by continuously checking the rank of the interaction matrix. At the moment that some DOF's are lost, it is possible to fix the parent object pose and track only the articulated part which normally needs only one or two DOF's. If the parent object motion is not considered, equation 6.6 takes the following form:

$$\mathbf{v}_2 = -\lambda \widehat{{}^C\mathbf{W}_O}\mathbf{S}^{\perp} \left(\widehat{\mathbf{D}}\widehat{\mathbf{L}_{s_2}} \widehat{{}^C\mathbf{W}_O}\mathbf{S}^{\perp} \right)^{+} \widehat{\mathbf{D}}(\mathbf{s}(\mathbf{c}_O^2) - \mathbf{s}_2^*) \tag{6.7}$$

Figures 6.4 and 6.5 show the case where there is not enough information for estimating the full 6D pose, but it is still possible to track the articulated part along one translational DOF. Even though the articulated DOF's can be successfully tracked, the initial error on the rest of DOF's cannot be corrected, because the parent object edges are not considered in the minimization.

6.2.2.4 Active Pose Estimation

One possibility to improve the set of visual features is to apply the VVS method simultaneously on several images taken from different point of views. For example, if the robot is too close to the target object so that it is not possible to obtain an image with enough visual features, it would be possible to move the robot vision system to different points of view, in order to obtain several images which provide the necessary information. An approach for performing VVS on a stereo system was proposed in [40]. We generalize their approach to any number of views, and consider an active vision framework, where the camera is moved in order to acquire suitable views. We have called this method the *active pose estimation*.

One important assumption of this approach is that the target object must be fixed in the environment, or either that its motion is known. However, this normally holds true for large objects found in household environments, like a cupboard, a fridge, the dishwasher, etc. During operation, it is not expected that these objects experiment any motion other than the internal motion of its articulated parts. In addition, it is also assumed that the camera displacement between views is known.

Active pose estimation can be built on the basis of the VVS method, by stacking the feature vectors and the interaction matrices obtained from each image, and performing the virtual control law in a common frame. Let ${}^{C_i}\mathbf{M}_{C_0}$ denote the pose of the initial camera position with respect to the i-th one, and ${}^{C_i}\mathbf{W}_{C_0}$ its corresponding twist transformation matrix. Let $\mathbf{s}^i(\mathbf{c}_O^i)$ be the set of visual features obtained from the i-th view, and $\widehat{\mathbf{L}_s^i}$ its associated interaction matrix. The global interaction matrix can be computed by stacking those defined for each image, as:

$$\widehat{\mathbf{L}_s} = \begin{pmatrix} \widehat{\mathbf{L}_s^0} & & \cdots & 0 \\ & \widehat{\mathbf{L}_s^1} & & \vdots \\ \vdots & & \ddots & \\ 0 & \cdots & & \widehat{\mathbf{L}_s^n} \end{pmatrix}$$

Similarly, the global features vector is built as $\mathbf{s} = \left(\mathbf{s}^0, \mathbf{s}^1, \ldots \mathbf{s}^n\right)^T$. The VVS control law for the set of i-th images can then be defined as:

$$\mathbf{v}_C^0 = -\lambda \left(\widehat{\mathbf{D}}\widehat{\mathbf{L}_s \mathbf{W}} \right)^{+} \widehat{\mathbf{D}}\left(\mathbf{s} - \mathbf{s}^*\right) \tag{6.8}$$

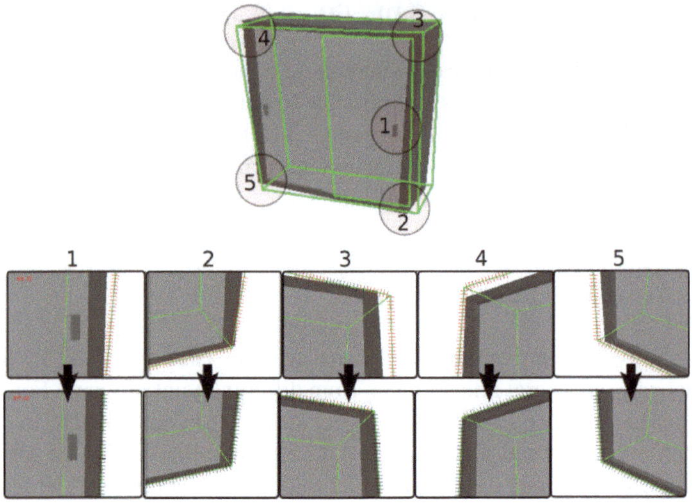

Fig. 6.6 Five different views of the cabinet object. The top row shows the projection of the edges according to the initial pose estimation. The bottom row shows the projection after applying the VVS method.

where $\widehat{\mathbf{W}} = \left({}^{C_0}\mathbf{W}_{C_0}, {}^{C_1}\mathbf{W}_{C_0}, \ldots, {}^{C_n}\mathbf{W}_{C_0}\right)^T$. $\mathbf{v}_C^0$ is then used to update the pose estimation $\mathbf{c}_O^0$, which is later propagated to the rest of views by means of the corresponding transformations ${}^{C_i}\mathbf{M}_{C_0}$.

As an experiment to show the benefits of active pose estimation with respect to standard approaches, we consider the problem of estimating the pose of the cabinet object used in the previous experiments, when the robot is placed in front of it. More concretely, we consider the camera to be at approximately one meter in front of the the cabinet, and looking towards the handle. At this distance, only a small part of the cabinet is visible on the image, as shown in the first point of view in Figure 6.6. As we have already shown in the previous sections, the image obtained from this point of view is not enough for ensuring the full rank of the interaction matrix. In order to obtain further visual features, the active pose estimation approach is applied by making the robot camera look towards the corners of the object, as shown in Figure 6.6.

Even though an important error has been set to the initial pose estimation, the active pose approach is able to converge in all the images. Figure 6.7 shows the evolution of the pose error with the active pose approach when using five images, in comparison with the standard pose estimation using only one image. As it is expected, the pose estimation with five images provides much more accuracy than the standard one-image approach, which, in this case, is not able to converge due to the weak feature set.

Figure 6.8 shows how the pose error norm decreases as more images are used, whereas Figure 6.9 compares the pose accuracy obtained by the standard method using one image taken from a far position, with respect to the active pose approach using five images.

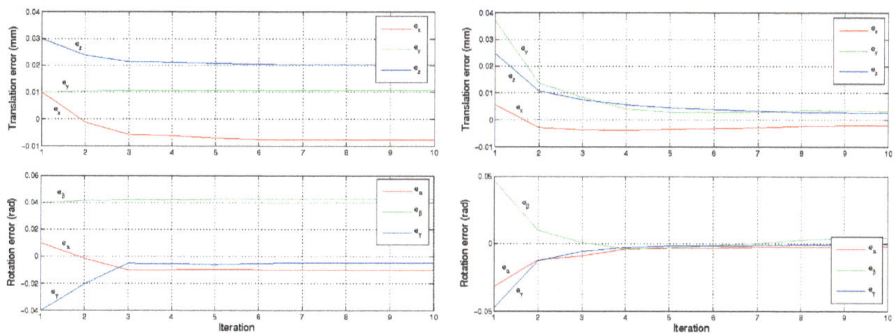

(a) Standard VVS-based pose estimation on the first view
(b) Active pose estimation with five views

Fig. 6.7 Comparison of the convergence of the pose components

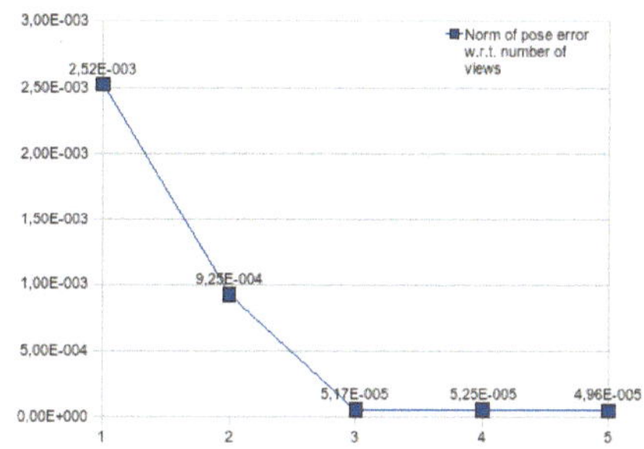

Fig. 6.8 Evolution of the euclidean norm of the pose error vector according to the number of views considered during the active pose estimation

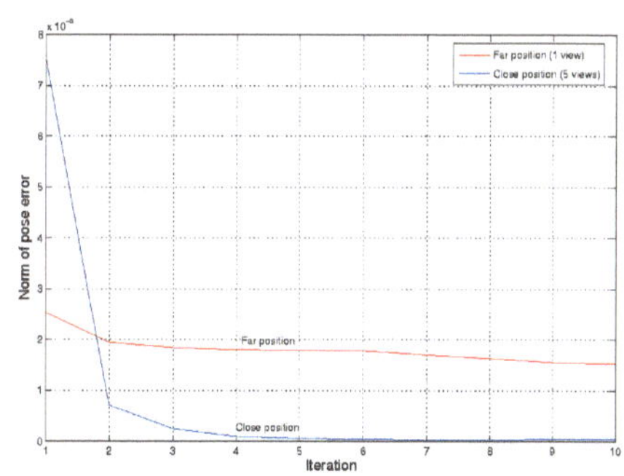

Fig. 6.9 Evolution of the norm of the pose error vector under standard VVS-based pose estimation from a far point of view, versus active pose estimation with five views

In conclusion, active pose estimation can provide an accurate pose in situations where the robot is close to the object and the other methods cannot be applied. By taking several images of the object, from different points of view, the number of visual features from which to build a full-rank interaction matrix is increased. However, pose errors can still appear, making it necessary to combine the visual information with force feedback. In the following section, the problem of vision-force control is addressed.

6.3 Vision-Force Control

As it has been shown in the previous section, vision can provide accurate 6D localization in cases where there is enough information in the image for ensuring a full rank interaction matrix. However, local minima and pose ambiguities can appear, leading to important errors in the object pose estimation. In addition, image processing algorithms can easily fail due to changing lighting conditions, occlusions, etc. At the moment of manipulation, vision errors are manifested in misalignments between the hand and the object, which can lead to failure in the physical interaction task. Therefore, it is desirable to combine the vision control signal with that coming from a force controller, as they allow to eventually correct the end-effector trajectory when contacts appear.

On the contrary, as described in section 5.2, position-force control approaches require, in general, geometric knowledge about the environment in order to perform the position-based trajectory required by the particular task. To overcome this deficiency, the use of a vision control loop instead of the position one has been investigated. Indeed, computer vision can provide a powerful way of sensing the environment and can potentially reduce or avoid the need for environmental modeling. Vision allows accurate part alignment in partially unknown and/or dynamic environments, before and during the contact phase.

When dealing with disparate sensors, the fundamental question of how to effectively combine their measurements must be addressed. One approach is to combine the measurements using multi-sensor fusion techniques [8]. However, such method is not suitable for vision and force sensors, since the data they provide measure fundamentally different physical phenomena. The second approach is to combine visual and force data at the control level.

For this, two main approaches have been studied: *impedance-based* [124] and *hybrid-based* control [70, 131, 7]. These approaches are equivalent to the position-force methods, with the only difference that the position controller is replaced by a vision one. Therefore, they share the same drawbacks than the corresponding position-force schemes. In this section we present a survey of the impedance and hybrid vision-force approaches, and propose a new vision-force control scheme, based on the concept of external control [36, 141], that allows to solve the deficiencies in current approaches.

Our purpose is not to provide a complete review of vision-force control approaches, but to show with representative examples the general structure of actual schemes, and therefore, to prove the originality of the new *external vision-force control* approach. We first present the hybrid vision-force method, and then we focus on the impedance-based approach. Finally, the new method is described and compared with respect to the existing ones.

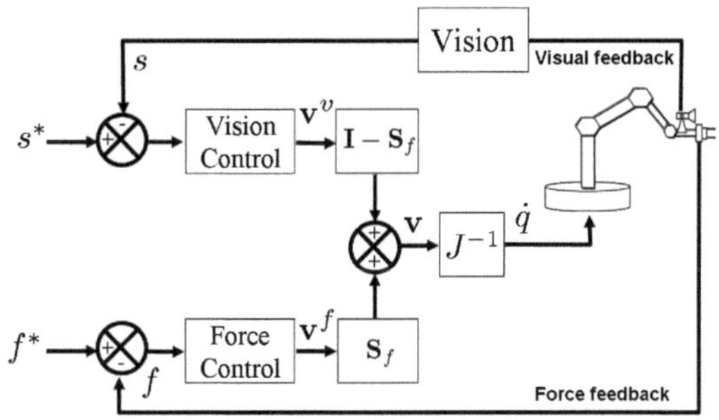

Fig. 6.10 Hybrid vision-force control approach

6.3.1 *Hybrid Vision-Force Control*

Hybrid vision-force control schemes follow the architecture shown in Figure 6.10, which is equivalent to hybrid position-force control, with the difference that the position controller is replaced by a vision controller and that the selection matrices are applied after the controllers, at the velocity level. The input to the vision control law is computed from the error between the desired and the current visual features, $\mathbf{s}^*$ and $\mathbf{s}$. In order to avoid any conflict at the actuator level, it is necessary to ensure orthogonality between the vision and the force controller outputs $\mathbf{v}^v$ and $\mathbf{v}^f$. For this, a diagonal selection matrix, $\mathbf{S_f}$, and its complementary, $\mathbf{I}-\mathbf{S_f}$, are introduced into both control loops. In order to produce a resulting motion, vision and force contributions are combined in the following way:

$$\mathbf{v} = (\mathbf{I}-\mathbf{S_f})\mathbf{v}^v + \mathbf{S_f}\mathbf{v}^f \tag{6.9}$$

where $\mathbf{v}^v$ is the output of the vision control law according to the general expression 6.4, and $\mathbf{v}^f$ is the output of the particular force control law.

Hybrid vision-force control has been successfully implemented, for example, in [133] to grasp electrical lines through tele-operation, in [131], where a manipulator is controlled in order to maintain a constant force, normal to a moving plate, while maintaining the point of contact in the center of the plate, or in [7], for contour following in industrial applications.

Hybrid control allows visual servoing only in those directions that are orthogonal to directions along which force feedback is used. If a perturbation occurs in the force controlled direction, the positioning error generated cannot be corrected by the vision-based controller.

In order to illustrate this problem, we consider a simulated peg-in-hole task, where a robot with an eye-in-hand camera has to insert a peg of 10 cm length into a hole of the same depth. The hole diameter is 3 mm bigger than the peg diameter. The hole is considered to be in the center of a pattern composed by four circles forming a square, as shown in Figure 6.11. We assume that the hole, and thus also the

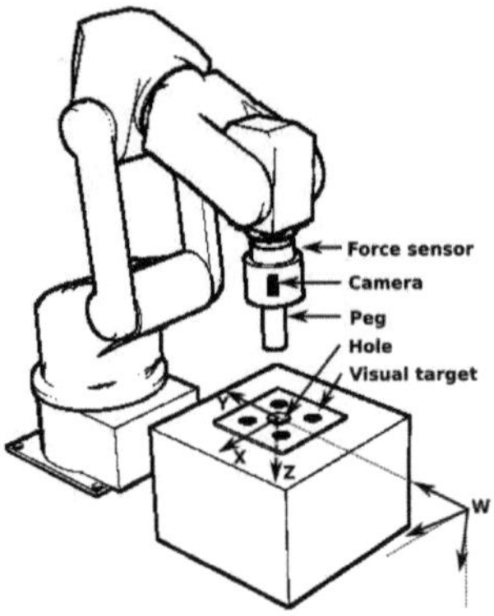

Fig. 6.11 General setup of the peg-in-hole vision-force-guided task

pattern, are on the origin of the world frame, but rotated 10 degrees around Y axis. The goal of the vision part is to reach a camera position so that the square is centered on the image at a given size (see Figure 6.12a). This desired position corresponds to the case when the peg is successfully inserted into the hole, and is learnt during an off-line process.

Simulation results of the task performance using a hybrid vision-force controller are shown in Figures 6.12 and 6.13. The vision controller implements an image-based approach using image point features, $\mathbf{s}_i = (u_i, v_i)^T$, whereas the force controller implements stiffness control. The force selection matrix is chosen to be the null matrix when there is no contact. Once contact is detected, the selection matrix is set to $\mathbf{S_f} = \mathbf{diag}(0,0,1,1,1,0)$ according to the task to perform. By setting the initial camera cartesian position to $\mathbf{c}_W = (0.02,-0.02,-0.36,-0.087,0.087,0)^T$, expressed in the world frame, the corresponding initial image features are shown in Figure 6.12b. Figures 6.13a and 6.13b show how the desired image position is never reached, and so, the image error does not converge to zero. This happens because this control law relies entirely on the selection matrix, which is task-dependent, and only has meaning when the tool frame is aligned with the compliance frame where $\mathbf{S_f}$ has been defined. For our chosen initial position, both frames are not aligned. The control law starts to converge as long as no contact is detected (see the dynamic and kinematic screw on the very first iterations in Figures 6.13c and 6.13d), because in this case all degrees of freedom are vision-controlled. The first contact appears on the surface than contains the hole, before the peg is completely aligned with it. Some degrees of freedom switch to force-control, and thus the vision error on these directions is not corrected.

In conclusion, the hybrid-based vision-force control is only valid when the task is perfectly known and the controlled frame is well-aligned with the natural constraints. If contacts appear before vision has converged, then the desired position may not be reached.

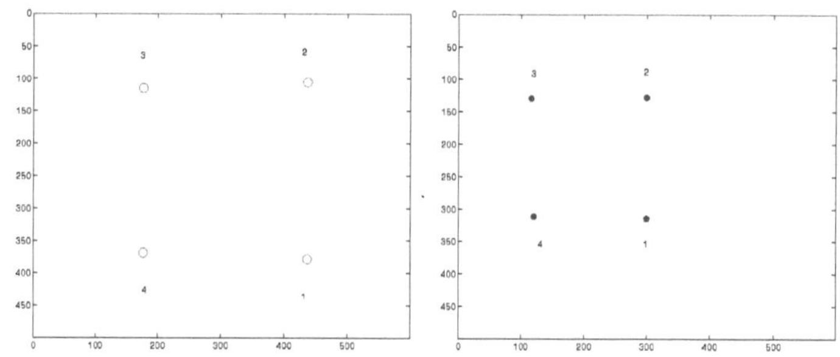

(a) Desired state in image coordinates (b) Initial state in image coordinates

Fig. 6.12 Visual features for the hybrid vision-force experiment.

6.3.2 Impedance Vision-Force Control

In the vision-force impedance control (see Figure 6.14), the final control vector can be written as $\mathbf{v} = \mathbf{v}^v + \mathbf{v}^f$, where $\mathbf{v}^v$ is the kinematic screw computed by the vision control law, and $\mathbf{v}^f$ is computed by the force control loop. By considering only the stiffness matrix in the impedance control law, and a generic visual servoing approach, the expression for impedance vision-force control can be written as:

$$\mathbf{v} = -\lambda \widehat{\mathbf{L}^{+}}(\mathbf{s}-\mathbf{s}^*) + \widehat{\mathbf{L}_\times}^{-1}\mathbf{K}^{-1}(\mathbf{f}-\mathbf{f}^*) \qquad (6.10)$$

where $\widehat{\mathbf{L}_\times}$ relates $\mathbf{v}$ and $\dot{\mathbf{x}}$ according to $\dot{\mathbf{x}} = \widehat{\mathbf{L}_\times} \cdot \mathbf{v}$ [113, 183].

With such a serial scheme, it is possible that $\mathbf{s} \neq \mathbf{s}^*$ and $\mathbf{f} \neq \mathbf{f}^*$ while $\mathbf{v} = 0$. A local minima can thus be attained and oscillatory phenomena can occur.

This is shown in simulation in Figures 6.15 and 6.16. Figure 6.15b shows the camera image when it is in the initial cartesian position $\mathbf{c}_W = (0,0,-0.25,0,0,1.57)$ with respect to the world frame, whereas Figure 6.15a shows the desired state. We can see in Figures 6.16a and 6.16b how the system starts to converge, but, finally, the desired position is never reached. We can also see in Figures 6.16c and 6.16d an oscillatory behavior of the dynamic and kinematic screw. This is because the control law has reached a local minima. The vision control law computes a kinematic screw for reducing the image error, but it is opposite to the dynamic screw computed by the force control law. As coupling is done in cartesian space, the result is that the robot is continuously oscillating between the direction that reduces the vision error and the one that reduces the force error.

6.3.3 External Hybrid Vision-Force Control

We have analyzed the existing methods for vision-force control and have shown their main drawbacks by simulation results. The hybrid vision-force approach is

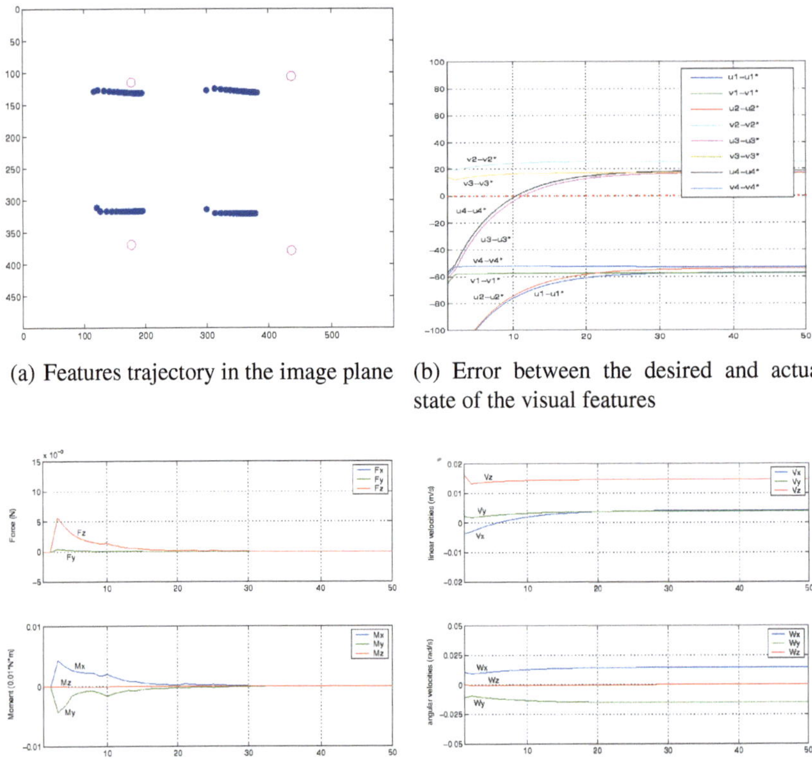

(a) Features trajectory in the image plane

(b) Error between the desired and actual state of the visual features

(c) Force signal

(d) Kinematic screw

Fig. 6.13 Hybrid vision-force results

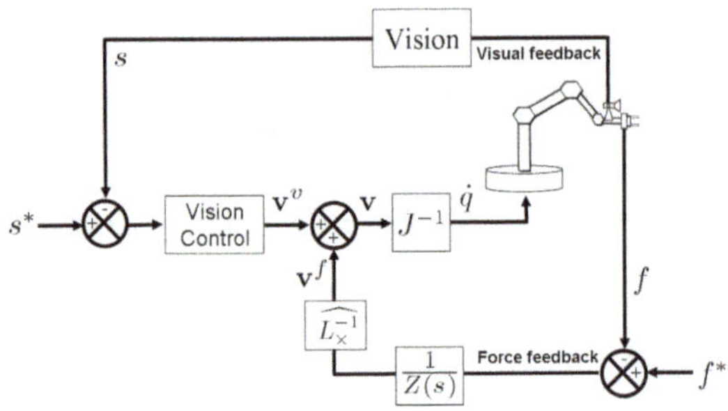

Fig. 6.14 Impedance vision-force control approach

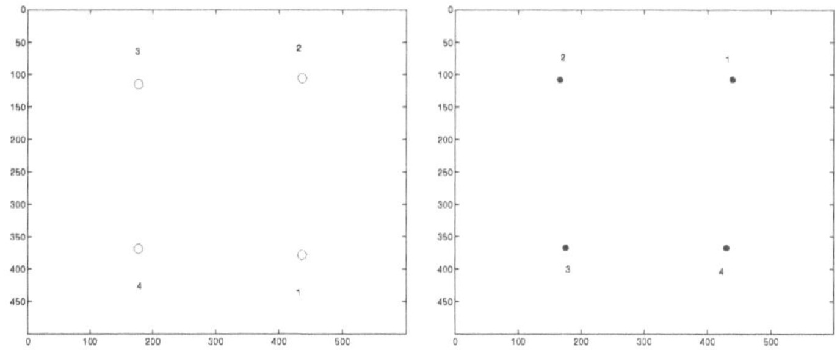

(a) Desired state in image coordinates (b) Initial state in image coordinates

Fig. 6.15 Visual features for the impedance vision-force experiment

unable to control all the degrees of freedom in vision and force simultaneously. This leads to some cases where convergence is not possible. Regarding impedance vision-force control, the coupling between the vision and the force control law is done in the cartesian space. Local minima can be reached during convergence, making it impossible to simultaneously reach the desired force and vision references. In this section, we present a new approach, based on the concept of external control, that improves the performance obtained by the existing methods.

6.3.3.1 General Concept

In the external vision-force control approach, the force control loop is closed around an internal vision control loop in a hierarchical way (see Figure 6.17). The reference trajectory $\mathbf{s}^*$, used as input of the vision-based controller, is modified according to the external force control loop. Then, instead of adding the force control output to the vision control output as classical approaches do, the force control signal is projected into sensor space by means of $\widehat{\mathbf{L}_s}$, and used to modify the desired vector of visual features, $\mathbf{s}^*$. Either if the end-effector is in contact with the environment or not, the robot is only controlled by the visual control law, and thus, the proposed control scheme has the same stability and convergence properties as the particular visual control law we choose [76].

6.3.3.2 A Concrete Control Law

Although any direct force control method could be used, we show here a particular example using stiffness control. In the control scheme shown in Figure 6.17, the desired wrench, $\mathbf{f}^*$, is added as input in the force feedback control loop, without affecting the control law stability [123]. The stiffness is controlled by the force controller according to a proportional control law:

$$d\mathbf{x} = {}^C\mathbf{W}_F\mathbf{K}^{-1}(\mathbf{f} - \mathbf{f}^*) \tag{6.11}$$

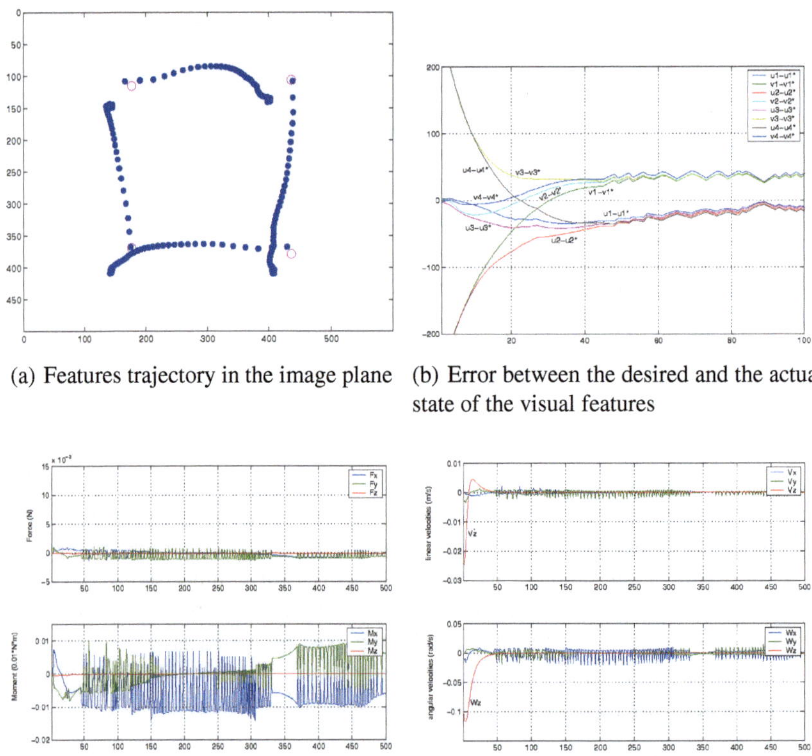

(a) Features trajectory in the image plane (b) Error between the desired and the actual state of the visual features

(c) Force signal (d) Kinematic screw

Fig. 6.16 Impedance vision-force results

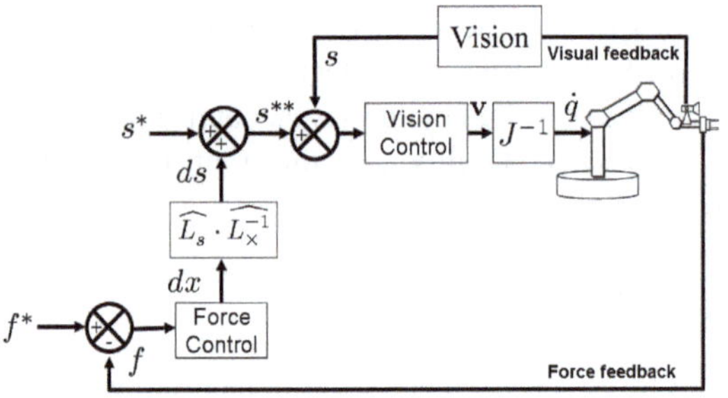

Fig. 6.17 External hybrid vision/force control loop

where ${}^{C}\mathbf{W}_F$ is the matrix that transforms screws from the force sensor frame to the vision control space. Unlike existing approaches, the force controller does not modify the vision control output. Instead, it only modifies the reference trajectory of visual observations, $\mathbf{s}^*$:

$$\mathbf{s}^{**} = \mathbf{s}^* + d\mathbf{s} \tag{6.12}$$

where $\mathbf{s}^{**}$ is the modified reference for visual features and $d\mathbf{s}$ can be computed by projecting $d\mathbf{x}$ by means of the interaction matrix, as $d\mathbf{s} = \widehat{\mathbf{L}_s} \cdot \widehat{\mathbf{L}_\times^{-1}} \cdot d\mathbf{x}$.

The hierarchical juxtaposition of the force control loop on the vision control loop provides several advantages with respect to the existing methods: first, selection matrices and time-dependent geometric transformations are eliminated from the control loop leading to a controller design independent of the environment configuration. Since the force control only acts on the reference trajectory, conflicts between force and vision controllers in the cartesian space are avoided. In addition, the external nature of the force control loop allows to easily implement this method over existing vision-based controllers.

6.3.3.3 Simulation Results

Figures 6.18 and 6.19 show how the external hybrid control succeeds in the situations where the hybrid-based and impedance-based control failed.

In the first example, as it is shown in Figures 6.18c and 6.18d, there is a force in Z direction (approaching direction) that corresponds to the contact with the surface before inserting the peg. The velocity becomes zero in this direction, but not in the others, meaning that, even in the presence of contact, vision is converging in the non-constrained directions. During this stage, the peg is in contact with the external surface, moving in a tangent direction until the hole is found, thanks to the vision convergence. At this point, the force in Z disappears and the peg is inserted successfully.

In the second example, where impedance-based control fails, the external approach succeeds, because it avoids the conflict between the vision and force control signals at the cartesian level, by hierarchically putting the force loop around the vision loop. The coupling is thus done in sensor space, and the control law that finally acts on the robot is only the vision one. Figure 6.19c shows that some torques appear during the execution of the task. These torques temporarily modify the vision reference and, thus, the vision control law computes a control vector that helps the convergence of both vision and force.

6.4 Vision-Force Suitability for Physical Interaction

In this section we discuss how the vision-force control strategies described in the previous points can be adopted in the different stages of a physical interaction task, i.e. during grasping and interaction.

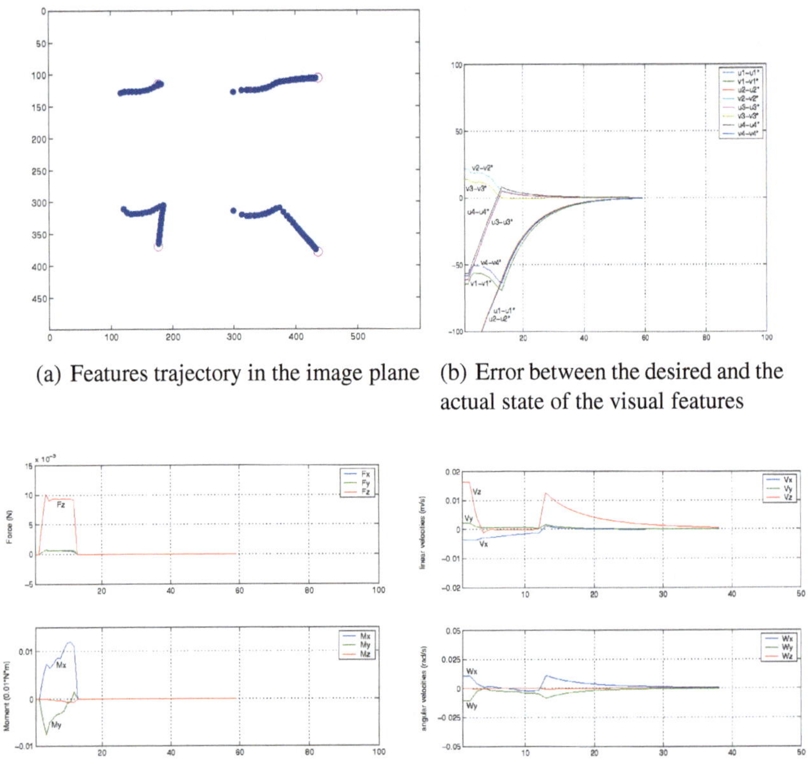

(a) Features trajectory in the image plane (b) Error between the desired and the actual state of the visual features

(c) Force signal (d) Kinematic screw

Fig. 6.18 External vision-force vs. Hybrid vision-force

6.4.1 Vision-Force Control for Grasping

Vision is the most important sensor during the reaching phase. When there is still no contact with the environment and, thus, force and tactile information is not available, vision-based techniques can be adopted in order to locate the target object in the environment and guide the hand towards it.

Pose estimation methods can provide an approximate position and orientation of the target object in the workspace, which is absolutely needed for starting the reach-to-grasp action. Depending on the adopted approach, more or less accuracy will be obtained in the pose estimation, and different kind of previous knowledge will be required. For example, an appearance-based pose estimation approach does not require knowledge about the object 3D model. However, pose accuracy is normally worse than in model-based methods, but these usually require a suitable initial estimation in addition to the object model. The task programmer has to chose the most suitable approach depending on the available previous knowledge.

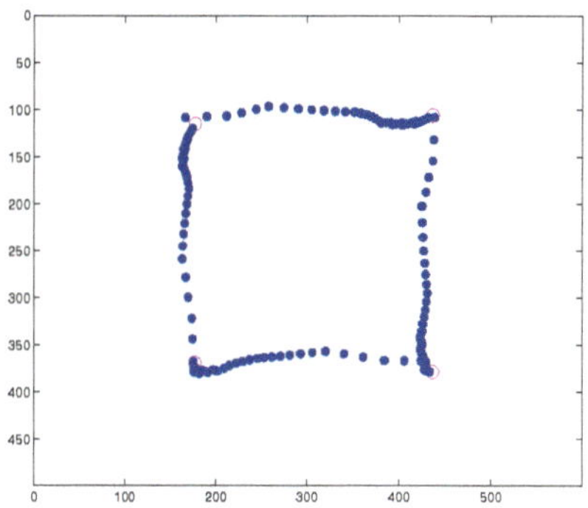

(a) Features trajectory in the image plane

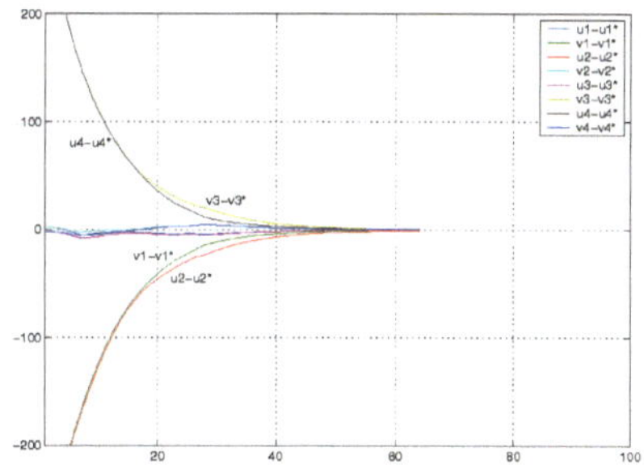

(b) Error between the desired and the actual state of the visual features

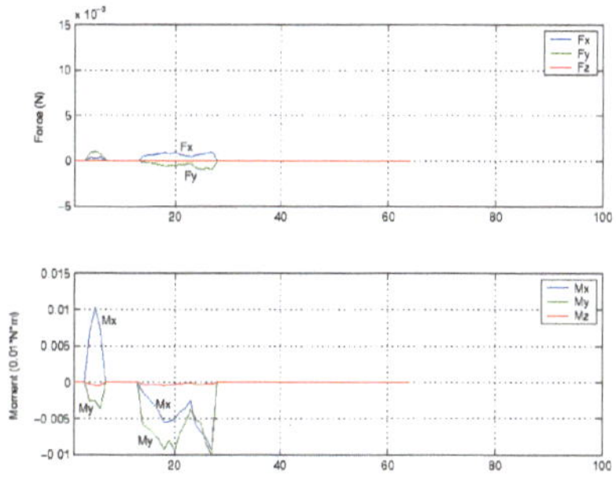

(c) Force signal

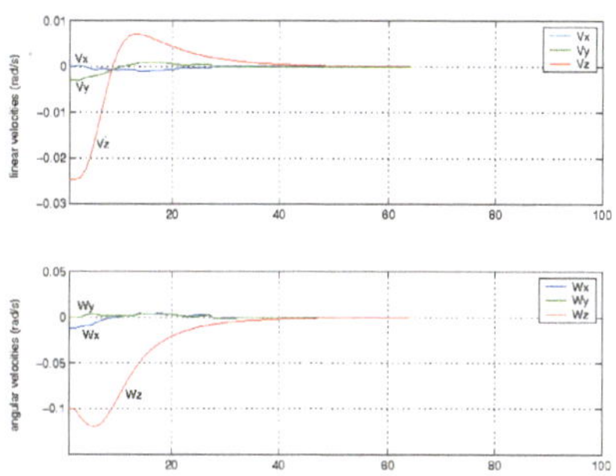

(d) Kinematic screw

Fig. 6.19 External vision-force vs. Impedance vision-force

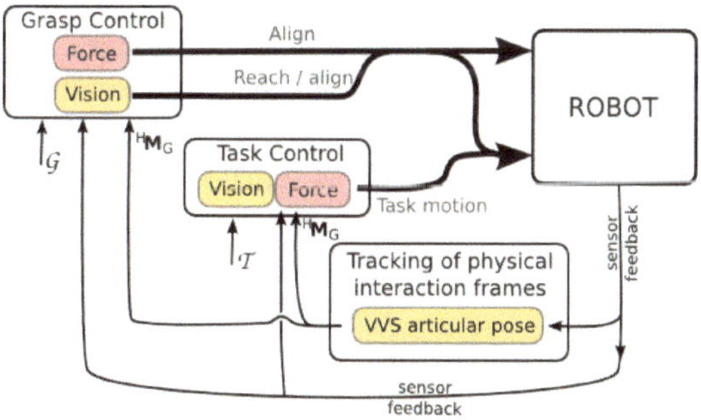

Fig. 6.20 The general vision-force control scheme in terms of the physical interaction framework

In the vision-based examples of this book, we assume that object models models have been previously acquired through an object recognition method. Therefore, we focus on model-based pose estimation techniques, and, more concretely, those explained in section 6.2.2.

VVS-based pose estimation is not only a method for estimating the object position and orientation, but also to track it under object or camera motion. Therefore, it is useful for tracking part of the physical interaction frames, concretely the object, task and grasp frames. The hand frame can also be tracked, either by the direct visual observation of the hand, or through the robot kinematics, in the case of a calibrated camera. Therefore, vision is the only sensor which can provide a direct estimation of the hand-grasp transformation during the non-contact phase. During interaction, this tracking can provide important information about the object mechanism, thus allowing the robot to adapt to the particular constrained trajectory.

In conclusion, as summarized in figure 6.20, the main contribution of vision to the grasping part of a physical interaction task is that it can potentially locate and track the objects in the environment, thus enabling reaching and accurate hand-object alignment, before and after contact is performed. However, after contact, it is desirable to perform the alignment task together with a force controller, in order to reliably cope with the hand-object misalignments generated by vision.

6.4.2 Vision-Force Control for Interaction

During interaction, the same vision-based pose estimation algorithms can be used in order to analyze the effects of the physical interaction on the object. When pulling open a door, for example, the door pose can be tracked while the robot performs the task motion, and important events can be detected, such as that the desired opening angle has been reached, or that the grasp has failed, etc.

However, the main contribution of vision during interaction, is that it is possible to track the evolution of the physical interaction frames by a direct observation of the hand and the object. This avoids the use of force-based tracking techniques. The object evolution can be observed directly, and, thus, the hand motion can be adapted accordingly to the particular mechanism, without planning a predefined trajectory. Again, it is necessary to combine the visual signal with force feedback, in order to deal with unexpected forces due to vision processing errors.

6.5 Practical Implementation: ISRC Mobile Manipulator

The theoretical developments of this chapter have been applied to a real robot. In this section, we consider a mobile manipulator which is observing simultaneously his hand and the object to manipulate, by using an external camera (thus following and end-point closed loop approach). The physical interaction framework and planner are used in order to specify the physical interaction task and to plan a suitable grasp on the object. The external vision-force control approach introduced in section 6.3.3 is used in order to, first, guide the robot hand towards the grasp position and, second, perform the task taking into account external forces. This work was developed during a short stay of three months at *Intelligent Systems Research Center* (ISRC) of Sungkyunkwan University in Suwon, South Korea.

The problem of hand-object alignment for grasping tasks has been already addressed by other authors. In [69], a visual servoing framework for aligning the end-effector with an object was presented. Instead of working in the euclidean space, visual servoing was done on the projective space by performing projective reconstruction with a stereo camera, thus avoiding the need for camera calibration. The desired gripper-to-object relationship was learnt during an off-line procedure. In [180], a position-based visual servoing approach was used on a humanoid robot in order to guide the hand towards the object. Hand pose was estimated by a kalman filter taking as input the stereo reconstruction of a set of LEDs attached on the robot hand.

As in [180], we also adopt a position-based visual servoing control law, because of the facilities that this approach offers for task specification. Instead of using a stereo camera and performing 3D reconstruction, we make use of a single camera and follow the VVS approach described in section 6.2.2 for pose estimation. The goal of the vision control loop is to align the gripper with respect to some part of the object (e.g. the handle). As the pose of the gripper and the object is estimated online, the relative position between both, i.e. the grasp link, can be computed at each iteration without the need of knowing the position of the camera with respect to the robot base. Therefore, the robot is still able to perform the task even in the presence of camera motion. Task execution is independent of camera position. No extrinsic camera parameters are needed, which makes the integration of this approach into other robotic systems very easy, and opens the door to best-view planning algorithms for head control. In addition, instead of learning the grasp position during an offline stage like in [69], we make use of the physical interaction planner of chapter 4, which autonomously computes a suitable grasp for the given task. Finally, and in contrast with existing approaches, visual servoing does not finish when the robot grasps the object. Instead, the external vision-force control approach is adopted in order to perform a given task on the object. Thus, vision-force control is not only used for hand-object alignment (reaching), but also for task execution and supervision (interaction).

6.5.1 Description of the System

We consider a mobile manipulator composed of an Amtec ultra light-weight arm with 7 DOF's mounted on an ActivMedia PowerBot mobile robot. The hand of the robot is a PowerCube parallel jaw gripper. This robot belongs to the Intelligent Systems Research Center (Sungkyunkwan University, South Korea), and is already endowed with recognition and navigation capabilities [82, 100], so that it is able to recognize the object to manipulate and to retrieve its geometrical and structural model from a database. The robot is able to move in front of the object by using navigation capabilities such as mapping, localization, obstacle avoidance, etc.

The ultimate goal is to interact with the different furniture and articulated objects that can be found in home environments, such as doors, windows, wardrobes, drawers, lights, etc. For this particular application, we have chosen a door opening task, because it is the most common task in home environments. The task starts when the mobile manipulator has navigated in front of the object that is going to be manipulated and has a view of both the object and its hand. Experimental results are presented with two different doors: a closet and a refrigerator door.

6.5.2 Vision-Force Control for Everyday Tasks

A position-based visual servoing closed-loop approach has been adopted for reaching the objects. A robot head observes both the gripper and the part to manipulate, and tries to achieve a relative position between both, like in [69, 180]. For grasping and interaction, the vision is coupled with force information following the external vision-force approach. This allows to minimize the external forces while the vision control law performs the hand-object alignment. The controller details and experimental results are presented in the following sections.

6.5.2.1 Control Approach

Concretely, the external hybrid vision-force control approach described in section 6.3.3 is adopted. The force controller implements an active stiffness approach, whereas the vision controller is a position-based visual servoing control law.

We set the following vector of visual features:

$$\mathbf{s} = \begin{pmatrix} {}^H\mathbf{t}_G \\ \mathbf{u}\theta \end{pmatrix}$$

where ${}^H\mathbf{t}_G$ is the translational part of the grasp link, ${}^H\mathbf{M}_G$, and $\mathbf{u}\theta$ is the axis/angle representation of ${}^H\mathbf{R}_G$. For a comprehensive description of the frames involved in the vision task, see Figure 6.21.

The grasp link, ${}^H\mathbf{M}_G$, which relates the hand and the manipulated part, is computed directly from the visual observation of the gripper and the object, according to the following expression:

$${}^H\mathbf{M}_G = \left({}^C\mathbf{M}_{EP} \cdot {}^E\mathbf{M}_{EP}^{-1} \cdot {}^E\mathbf{M}_H\right)^{-1} \cdot {}^C\mathbf{M}_{OP} \cdot {}^O\mathbf{M}_{OP}^{-1} \cdot {}^O\mathbf{M}_G \tag{6.13}$$

where ${}^C\mathbf{M}_{EP}$ is an estimation of the pose of an arbitrary end-effector-related frame, expressed in the camera frame, and ${}^C\mathbf{M}_{OP}$ is an estimation of an arbitrary object frame pose, also expressed in the camera frame. In this application, the hand and the object pose is obtained by the VVS method, using a set of point features drawn

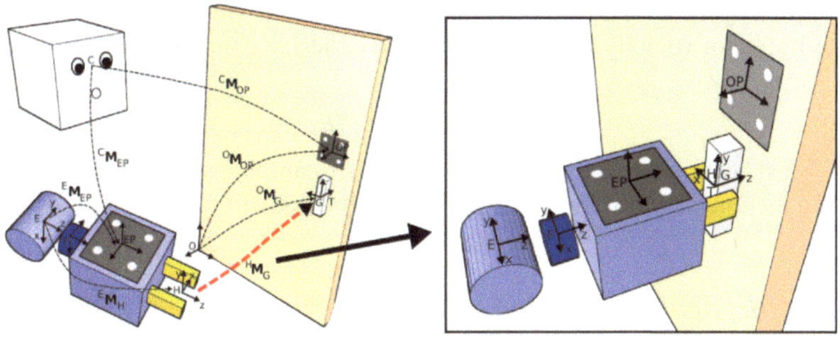

Fig. 6.21 The vision task is to establish and keep a suitable grasp link, i.e. aligning the hand frame with the grasp frame

on a pattern whose model is known. One pattern is attached to the gripper, at a known position ${}^{E}\mathbf{M}_{EP}$, and another pattern is attached to the object, also at a known position with respect to the object reference frame, ${}^{O}\mathbf{M}_{OP}$. In the next chapter, we will make use of the point-to-distance feature in order to use the natural edges of the object instead of the markers. Note that this does not significantly affect the current implementation, as VVS pose estimation can deal with different types of visual features [111].

Finally, the hand frame, ${}^{E}\mathbf{M}_{H}$, and the grasp frame, ${}^{O}\mathbf{M}_{G}$, together with a suitable task-oriented hand preshape, are computed by the physical interaction planner presented in chapter 4.

We then define $\mathbf{e}(\mathbf{s},\mathbf{s}^{**}) = \mathbf{s} - \mathbf{s}^{**}$ and compute the velocity in the hand frame, $\mathbf{v}_H$, using the classical approach:

$$\mathbf{v}_H = -\lambda \widehat{\mathbf{L}_s^{+}} \mathbf{e} + \widehat{\frac{\partial \mathbf{e}}{\partial t}} \tag{6.14}$$

$\mathbf{s}^{**}$ is computed from the force output according to expression 6.12, taking $\mathbf{s}^* = 0$, since the desired grasp link is ${}^{H}\mathbf{M}_G = \mathbf{I}$, which corresponds to a zero relative displacement. The last term, $\widehat{\frac{\partial \mathbf{e}}{\partial t}}$, is the estimation of the visual features motion which does not directly depend on the camera motion, and should be taken into account when the hand is in contact with the environment in order to reduce tracking errors. However, we can neglect it, because the use of force feedback allows us to cope with these small tracking errors, as long as the task velocity is small. The interaction matrix, $\widehat{\mathbf{L}_s}$, is set for the particular case of position-based visual servoing [23]:

$$\widehat{\mathbf{L}_s} = \begin{pmatrix} -\mathbf{I}_{3\times3} & \mathbf{0}_{3\times3} \\ \mathbf{0}_{3\times3} & -\mathbf{L}_w \end{pmatrix} \tag{6.15}$$

$$\mathbf{L}_w = \mathbf{I}_{3\times3} - \frac{\theta}{2}[\mathbf{u}]_\times + \left(1 - \frac{\operatorname{sinc}(\theta)}{\operatorname{sinc}^2(\frac{\theta}{2})}\right)[\mathbf{u}]_\times^2 \tag{6.16}$$

where $[\mathbf{u}]_\times$ is the skew-symmetric matrix associated to the rotation axis $\mathbf{u}$. Finally, the end-effector velocity which is sent to the cartesian controller is computed as:

$$\mathbf{v}_E = {}^E\mathbf{W}_H \cdot \mathbf{v}_H \tag{6.17}$$

In the next section we show how this controller can be used for both the grasping and interaction stages of the physical interaction task.

6.5.3 Physical Interaction

Experimental results are presented with two different doors: a closet and a refrigerator door. Both doors are very different in size, and have different handles. But both tasks can be described in terms of object actions, as the *Push backwards* action on a *Fixed handle* object. The physical interaction planning algorithm of chapter 4 selects a pinch precision task-oriented preshape and sets the physical interaction frames as shown in figure 6.21. The task frame is placed at the handle, thus allowing to specify the interaction task as pushing along negative Z direction, without requiring knowledge about the door hinge axis.

The whole manipulation process is divided into two steps: reaching the handle with a task-suitable grasp, and performing the task through interaction with the environment.

6.5.3.1 Reaching

The reaching task is divided into three different subtasks: reaching a pre-grasp position, reaching the grasp position and performing the grasp. The robot switches from one subtask to another when the resulting velocity of the vision-force controller is close to zero (i.e, the desired references have been reached).

First, a pre-grasp frame, P, is computed from the grasp frame and related to the object reference frame with the transformation ${}^O\mathbf{M}_P$. The pre-grasp position is used in order to adopt an initial configuration with respect to the final grasp frame so that the robot can reach the handle from a good direction.

The transformation between the hand frame and the pre-grasp frame, ${}^H\mathbf{M}_P$, is computed at each iteration by equation 6.13 (by replacing G with P), and then used for building the visual features vector, $\mathbf{s}$. During this step there is no contact with the environment. Thus, the force loop in the vision-force control law is not modifying the visual reference, and the system behaves purely according to the vision control law of expression 6.14.

Figure 6.22 shows the evolution of the visual velocity for each degree of freedom, in the case of the closet door. Initially, the hand frame is far from the pre-grasp frame (see Figure 6.23a), so that there is a large visual error. The visual control law makes this error converge to zero, which corresponds to the situation where the tool frame matches with the pre-grasp frame (see Figure 6.23b).

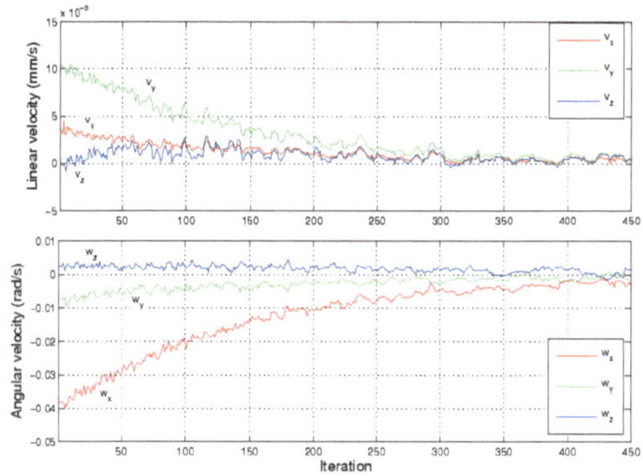

Fig. 6.22 Kinematic screw given in the end-effector frame, $\mathbf{v}_E$, for reaching the pre-grasp position in the closet door opening task

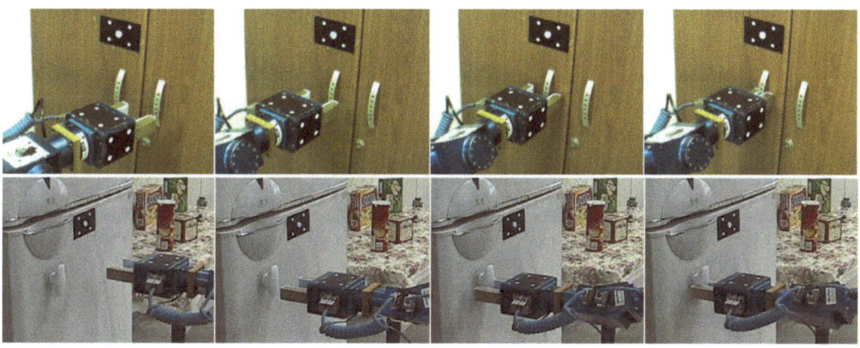

(a) Initial position (b) Reaching the pre-grasp position (c) Reaching the grasp position (d) Grasping

Fig. 6.23 Reaching the handle. Top row: closet. Bottom row: refrigerator.

After reaching the pre-grasp position, the hand frame is moved towards the grasp frame, as shown in Figure 6.23c. A new vector of visual features is computed according to the current position of the physical interaction frames, tracked by equation 6.13. Thus, the handle is reached from the reaching direction. At the end of this step, the grasp frame and the hand frame reach the desired relative transformation, thus establishing the grasp link. At this moment, the handle is situated between the robot fingertips.

The last step of the reaching stage is to grasp the handle. The previous step guarantees that the handle is between the robot fingertips. Then, the robot gripper is closed in order to grasp it, as shown in Figure 6.23d.

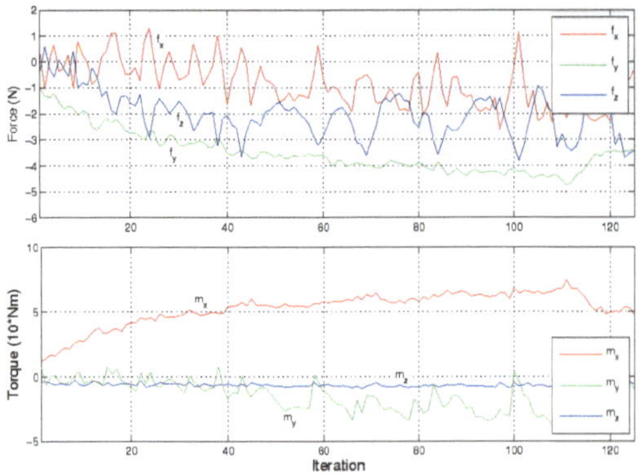

Fig. 6.24 Forces during grasping the closet handle, given in the force sensor frame, $\mathbf{f}_F$

During this step the first contacts appear. Thus, at the same time that the gripper is closed, the vision-force control law is active. The reference for the vision control law is to keep the grasp link (i.e, keep the handle in the middle point between both fingertips). The reference for the force control law is to minimize external forces, i.e. $\mathbf{f}_T^* = 0$. If, due to vision or modeling errors, the handle is not perfectly placed in the middle point between the fingertips, then one finger will make contact before the other. This will generate a force that the force control law will try to regulate to zero by modifying the vision reference. During this step, the stiffness coefficient on Y direction of the force sensor frame, F, is set to a small value in order to make the robot highly compliant in this direction.

As an example, Figure 6.24 shows the forces that appear in a particular experiment during the grasp of the closet door handle. Due to a premature contact on one of the fingers, a force in Y direction and a torque around X axis appear, expressed in the force sensor frame, F. These forces modify the visual reference (i.e. the desired grasp frame pose w.r.t the hand frame) according to expressions 6.11 and 6.12, and then the robot is visually-guided in order to reduce the force. Note that it would be impossible to correct this visual positioning error without using force feedback.

6.5.3.2 Interaction

Once the handle has been reached, the robot transforms the task specification, from the task frame to the end-effector frame, passing through the grasp link, according to the general expression 3.6 defined by the physical interaction framework. During this process, the vision-force controller is also active, and used for constantly monitoring and improving the grasp link.

The natural object mechanism will generate forces on the robot hand that the grasp controller will try to minimize, making the robot hand adapt to the particular constrained trajectory. Simultaneously, as the hand-object relative pose is observed

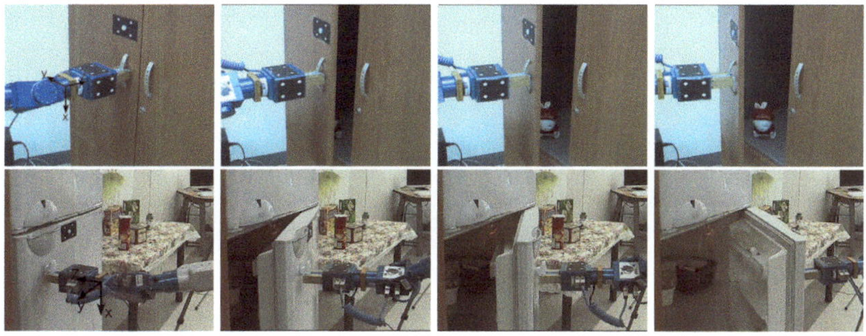

Fig. 6.25 Interaction phase. Top row: closet door. Bottom row: refrigerator door.

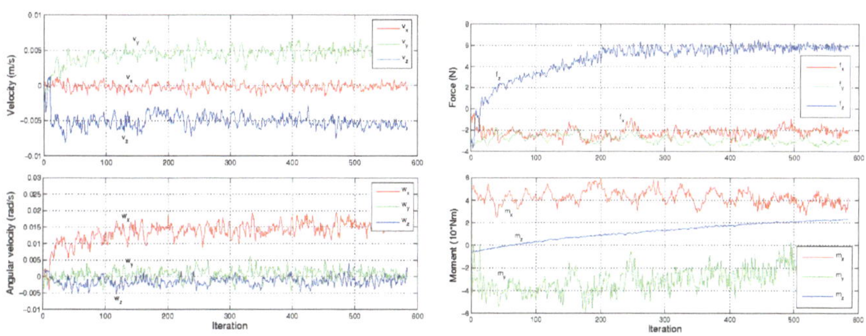

(a) Kinematic screw (b) Force signal

Fig. 6.26 Results of the external vision-force control law during the interaction phase with the closet

at each iteration, any misalignment between the hand and the handle will be also detected and corrected by the vision loop. Thus, both force and vision will work simultaneously for a common goal: performing the task while keeping a suitable grasp link.

Experimental results of the interaction phase can be seen in Figure 6.26 for the closet door, and in Figure 6.27 for the fridge. Concretely, figure 6.26b shows the evolution of the task forces given in the force sensor frame, $\mathbf{f}_F$, during the door opening action. These forces modify the visual reference, mainly in translation in Y and Z axis, and in rotation in X axis. The force in Z direction corresponds to the resistance of the particular object mechanism and is the one which is really acting on the task direction. The rest of forces appear on constrained directions and must be regulated to zero. The force in Y direction and torque in X direction are generated by the particular constrained trajectory when opening the door. The force control law updates the vision reference so that the robot hand adapts to the natural trajectory (see the velocity along Z and Y axis and the rotational velocity about X axis in Figure 6.26a).

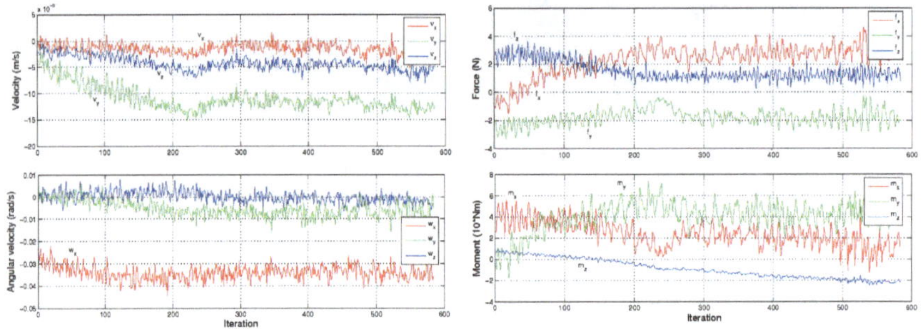

(a) Kinematic screw (b) Force signal

Fig. 6.27 Results of the external vision-force control law during the interaction phase with the fridge

It is worth noting that the hand trajectory is never planned. Instead, the vision-force control law adapts the hand motion automatically to the particular object mechanism. This is demonstrated by performing the same task specification on a fridge door where the hinge axis is at a different position. In fact, whereas the closet door opens to the left, the fridge does to the right. Consequently, angular velocity around X axis is negative in this case, as shown in Figure 6.27a.

6.6 Discussion and Conclusions

In this chapter we have shown how vision can assist force-based physical interaction. Vision can provide important information about the environment, which is impossible to obtain with contact-based sensors. Concretely, the position and orientation of objects in the environment is probably the most important information that manipulation algorithms require. Among the existing methods for vision-based pose recovery, we have focused on Virtual Visual Servoing, mainly because it is based on a solid background theory, and because it allows to track articulated structures efficiently.

However, convergence problems can appear when the number and nature of the visual features in the image is not enough for building a full-rank interaction matrix. This part could be improved by considering a wide-angle camera, or additional visual features, apart from the point-to-line distance.

Another solution is to adopt an active VVS pose estimation approach, which is one of the main contributions of this chapter. The concept is to acquire several views of the object and build a global interaction matrix from the visual features extracted from all of them. This method allows to recover the object pose in situations where the other approaches cannot be applied, e.g. when the robot camera is so close to the object that only a small part of it is visible, which is a common case in manipulation environments. It is out of the scope of this work to define best-view planning algorithms for the active VVS pose estimation method, but, in general, good views should include the object corners, or object parts where several edges converge.

For tasks that involve interaction with the environment, it is necessary to complement the visual information with force feedback by means of a vision-force control law. Previous work has been based on hybrid and impedance vision-force control, but, as we have shown in this chapter, both methods present problems that may affect the convergence. The second main contribution of this chapter is a new scheme for hybrid vision-force control based on the concept of external control. The hierarchical juxtaposition of the force control loop on the vision control loop provides several advantages with respect to the existing methods: selection matrices and time-dependent geometric transformations are eliminated from the control loop leading to a controller design independent of the environment configuration. Since the force control only acts on the vision reference trajectory, conflicts between force and vision controller at the cartesian level are avoided. In addition, the new method can be easily implemented on top of existing vision-based controllers. Although we have applied this control law with the particular case of position-based visual servoing, it can also be used with other kinds of visual servoing such as image-based or hybrid, as long as a suitable interaction matrix is computed.

A practical implementation concerning a mobile manipulator performing door opening tasks has been presented. The system is based on the physical interaction framework, and implements advanced vision-force servoing capabilities. The physical interaction planner computes a grasp on the handle, taking into account the task to perform, and an external position-based visual servoing approach is used in order to perform both the reaching and interaction phases of the physical interaction task.

Instead of putting the camera on the hand, which may cause some problems of visibility when the hand is close to the object, an external camera has been used, which allows to have a suitable view of the object even when contact is made. In addition, this is the typical configuration in humanoid robots, where the camera is placed on the head. An external camera also allows to visually track the robot hand pose and to estimate directly the grasp link. Tracking the hand could be avoided by using joint encoders to get the hand pose with respect to the camera, assuming that the pose of the camera with respect to the robot base is known, which is a difficult calibration problem (specially when the camera is not fixed). In practice, modeling errors would generate important errors in the hand pose estimation, making this approach unfeasible. It is for this reason that we compute simultaneously the object and the hand pose and work with the relative pose between both. The main advantage is that the camera can be moved freely without affecting the task execution. No external calibration is needed. The relative pose between the hand and the grasp frame is computed at each iteration, independently of the camera position, which makes our approach amenable to be integrated into current humanoid robots without hand-eye calibration. Finally, the task is executed by means of the novel external vision-force control approach. Both vision and force feedback cooperate during the task execution in order to keep the relative pose between the hand and the object, at the same time that the natural object mechanism is tracked. A natural extension of this work is to take advantage of the independence between task execution and camera motion, by developing head control algorithms that move the robot eyes, according to some optimization criteria, so that a suitable view of the hand and the object is always available.

Regarding the visual servoing control law, the position-based approach was chosen for our experiments, mainly because it works on the cartesian space, where the grasp and task are also defined, making it easier to generate the visual references from the grasp and task planning algorithms. However, it would be also possible to control the hand trajectory in cartesian space even with image-based visual servoing [118]. During the interaction phase, the robot is applying a motion on the object, and, thus, the visual features are in motion. We are currently neglecting the term that models this motion (see equation 6.14), and, therefore we have a tracking error, although force control can deal with it for small task velocities. It is worth noting that, due to image acquisition and processing times, the vision control frequency will be usually much smaller than the force control frequency. Thus, it is desired to run the global control law at the force sensor rate, even if the visual features are not updated at this high frequency. With this, we give priority to the force sensor feedback, and are able to detect and regulate contact at force sensor frequency, independently of the vision rate, which can vary from 25 Hz for ordinary cameras, up to 1 KHz for high-speed cameras.

In conclusion, the advantages that visual feedback can provide to force-based physical interaction have been shown in this chapter. Vision can be mainly used in order to estimate the position and orientation of objects in the environment, and, thus, it can provide an approximation of the location of the physical interaction frames. However, visual information can be subject to calibration errors, image processing problems, outliers, etc. In addition, there are situations in which an object pose cannot be estimated accurately, concretely when the image does not contain a rich set of visual features. Although these situations can be normally addressed through force control, in some cases force sensors still do not provide enough information for controlling some DOF's. This problem is addressed in the next chapter by the addition of tactile sensors, leading to vision-tactile-force control.

Chapter 7
Physical Interaction: The Contribution of Tactile Sensors

7.1 Brief Introduction

Touch is the ability to sense at the finger-object level [71]. It is known that there are about 17000 mechanoreceptors distributed along the fingers and the palm of the human hand, which provide rich information, mainly about the contact distribution, limb motion and forces. They can be classified into fast adapting (FA) and slowly adapting (SA) mechanoreceptors, being the former suitable for measuring skin vibrations, and the latter able to detect static stimuli. Tactile feedback is used by humans for retrieving object information, such as texture, temperature, shape, etc. and to assist grasping and manipulation with the detection of important events, like slip, or object deformation. By monitoring these events, the human hand is able to update the grasping force according the object weight, stiffness, friction, etc. In fact, it has been shown that people have difficulties to perform manipulation tasks when deprived of tactile feedback [83].

A tactile sensor is a device that can measure a given property of an object through physical contact between the sensor and the object [99]. It can be used to detect whether we are in contact, the contact configuration, slip, and even the object geometry if used with an active sensing approach. This valuable information can be used inside a control scheme, in order to regulate the grasping force, reach a desired contact configuration, etc.

It is worth mentioning that *tactile sensing* and *contact sensing* refer to different concepts. Tactile sensing includes research on skin-like sensors to measure pressure distribution, temperature, softness, etc., whereas contact sensing refers to the perception of forces and torques generated through the contact points during manipulation. This chapter is focused on the integration of tactile sensing with vision and force sensing, but forces will be considered at the wrist level, and not at the contact level.

Several attempts to mimic the human sense of touch exist [181], using different tactile technology such as conductive elastometers, elastomer-dielectric capacitive, light-based sensors, organic field-effect transistors, piezoelectric, or force sensing

M. Prats et al.: Robot Physical Interaction through the Combination of Vision, STAR 84, pp. 111–128.
springerlink.com

resistors, among others [72]. Unfortunately, each type of sensor is focused to a particular application, and it is still not possible to achieve the versatility of the human receptors.

Whereas vision and force integration techniques have been extensively studied in the robotics literature, tactile information has been rarely considered jointly with those sensors. Most of the applications of the tactile technology in robotics are found in minimally invasive surgery systems [49], where the human is still in the feedback loop. Very few approaches have considered the use of tactile feedback inside autonomous control schemes.

In this chapter, we study how tactile feedback can assist the execution of physical interaction tasks. For this, the information obtained from tactile array sensors is fused with force and vision feedback, providing the robot with a rich sensory experience which is then used for performing physical interaction tasks in a robust manner.

We propose a novel approach for the integration of these sensors, where vision and tactile information is fused first, together with object-robot localization information, and then used as input for a stiffness force controller. The vision controller implements the methods described in the previous chapter, and controls up to 6 cartesian DOF's. The tactile controller is in charge of three cartesian DOF's, allowing to keep always a good contact. Force control is performed through an active stiffness approach with a explicit force reference in the loop. The redundant nature of the sensors allows to perform the task even if a sensor is not available or provides inaccurate data.

The purpose of this chapter is to show the advantages that tactile sensing can provide to force-based and vision-force-based physical interaction. First, a bibliography review of tactile-based control approaches is introduced in section 7.2, and our approach to vision-tactile-force control is described in section 7.3. We then analyze how vision-force-tactile information can be used in terms of the physical interaction framework, and show a practical example of a door opening task using these sensors. Several experiments are performed, first by considering only force feedback, and then adding vision and tactile information. Robot-object localization errors are manually introduced in the experiments, and results show how the vision-tactile-force combination is able to deal with them, performing much better than the vision-force and force-only approaches. Although the proposed approach also supports vision-tactile integration, it has not been considered in the experiments, because, as described in chapter 5, we support that force sensing is the basic sensor for manipulation and it must be always present.

7.2 Tactile Control

In the robotics literature, tactile feedback has been used mainly for event-driven manipulation [71], which deals with switching between different task states according to a set of sensor-based conditions. Among the interesting events that can be detected with tactile sensors, it is worth mentioning perceiving a contact, an incipient slip, the contact type, changes in the object properties, etc. For example, in [32] force and tactile-based events were used for switching between different manipulation phases. The works in [12] and [34] also focused on tactile-based event detection, particularly slip and changes in the object texture. In [145] tactile feedback was adopted in order to recover the position and orientation of an object, following an active sensing approach.

Tactile information has been rarely used inside a control loop, mainly because of the noisy signals that tactile sensors provide. Most of the works that follow this approach are aimed at controlling the grasping force for avoiding slip [117, 80, 61], although other applications exist, such as that in [137], where whole-body contact feedback is used to lift a 30Kg box with a humanoid robot.

Due to the local nature of the information that tactile sensors provide, it is common to use them in combination with force and vision sensors. Vision-tactile-force combination was already addressed in [2], where some guidelines for detecting useful manipulation events with these sensors were given, without addressing the robot control problem. In [173], an approach for combining vision and tactile feedback in a control law was proposed, where force along one direction was also considered, and measured from the tactile sensors. A recent paper, [165], makes a comparison between tactile-only, force-only and force-tactile integration in the task of opening a door, where vision is used in a previous step in order to estimate the door handle pose.

We propose a new approach that differs from the previous ones in the following aspects:

- First, instead of an ad-hoc vision processing algorithm, we make use of the methods described in chapter 6, and more concretely, the VVS approach for model-based pose estimation and tracking of articulated objects. This makes our approach amenable to be used for different objects and tasks with little modification of the vision part, and provides a robust estimation, based on a well-established theory.
- Second, we consider a dedicated force-torque sensor providing 6 degrees of freedom, instead of a one-dimensional force computed from the tactile sensor (tactile-based force) or hand strain gauge.
- Finally, we consider full vision-tactile-force combination, in the sense that all the three sensors can be present at the same time. In addition, sensor combinations such as vision-force, tactile-force and vision-tactile are also supported.

7.3 Our Approach to Vision-Tactile-Force Control

In this chapter, we propose a position-vision-tactile hybrid control modified at the low-level by a force controller, as shown in Figure 7.1. In contrast to classical sensor fusion approaches based on kalman filters, hybrid control performs the integration of the sensor signals at the control level, which is more suitable in cases dealing

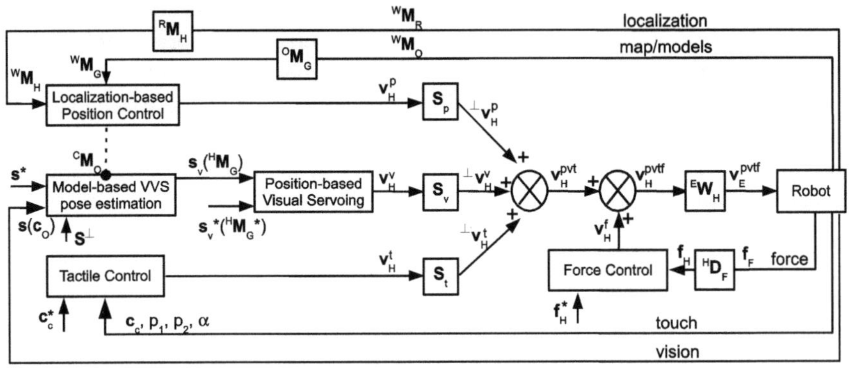

Fig. 7.1 Our control approach, integrating position, vision, tactile and force feedback

with disparate sensors [131]. In fact, vision, tactile and force sensors produce fundamentally different quantities, whereas traditional sensor fusion techniques require a common representation among the various sensors being integrated.

In the context of this controller, position control is understood as open-loop motion based on environment information obtained by mobile robot localization algorithms, either based on laser, sonar, odometry, an intelligent environment, etc. According to the hybrid control concept, only one sensor between localization, vision and tactile sensors is used for a given direction. Our approach is to use the one which provides the most accurate and robust information for that direction.

Thus, we establish a sensor hierarchy where tactile information is preferred over vision feedback, which is also preferred over localization information. The reason is that tactile sensors provide the most robust and detailed information about the object position, although at the contact level, whereas vision provides more global, but less accurate data, and localization is normally the most inaccurate source.

The cartesian DOF's assigned to each sensor are set on-line by three selection matrices, $\mathbf{S}_p$, $\mathbf{S}_v$ and $\mathbf{S}_t$, which must be mutually orthogonal, i.e. $\mathbf{S}_p \perp \mathbf{S}_v \perp \mathbf{S}_t \perp \mathbf{S}_p$. If a given cartesian direction must be explicitly controlled by force, it can be set to 0 on all the selection matrices, so that the force controller will fully take charge of it. However, it is worth noting that all the cartesian DOF's are under the effects of a force controller, even if there is no explicit reference for them. This allows to robustly perform physical interaction tasks even if visual or tactile information is not available.

Being $\mathbf{v}_H^p$, $\mathbf{v}_H^v$ and $\mathbf{v}_H^t$, the control velocity computed respectively by the position controller, the vision controller, and the tactile controller, all of them given in the hand frame, H, then the result of the preliminary sensor integration is given by:

$$\mathbf{v}_H^{pvt} = \mathbf{S}_p \cdot \mathbf{v}_H^p + \mathbf{S}_v \cdot \mathbf{v}_H^v + \mathbf{S}_t \cdot \mathbf{v}_H^t \tag{7.1}$$

It is worth mentioning that, in our approach, the selection matrices act on the control velocities, and not on the input errors as in the original hybrid control concept.

This is because the tactile and vision errors are not necessarily defined in the cartesian space, and thus, the selection matrices cannot be applied directly on them. Instead, they are applied after the corresponding controllers, where all the control signals are given in a common frame. Note that this is the common practice in hybrid vision-force control approaches [130]. This control velocity is then modified by a force controller which acts on all the degrees of freedom, ensuring that any force generated by a misalignment of the controlled frame, H, with respect to the environment will be kept inside a given range. If $\mathbf{v}_H^f$ is the hand velocity computed by the force controller, the final velocity signal, given in the robot end-effector frame can be computed as:

$$\mathbf{v}_E^{pvtf} = {}^E\mathbf{W}_H \cdot \left(\mathbf{v}_H^{pvt} + \mathbf{v}_H^f\right) \tag{7.2}$$

where ${}^E\mathbf{W}_H$ is the twist transformation matrix between the hand frame H, and the end-effector frame, E. A suitable cartesian controller, as that described in section 5.4.3, must then transform cartesian velocities into joint velocities.

This approach leads to a very natural behavior, where force is the most important sense, followed by tactile, vision, and localization sensors, in this order. Under a blind situation, the task can still be performed by position-tactile-force integration. If tactile feedback is not available, as for example in the phase of reaching an object, then position-vision-force can successfully guide the hand. In the worst case where tactile and vision are unavailable, position-force can still be used.

In the following points, the particular position, vision, tactile and force controllers used in our experiments are described. Although we propose here some specific controllers, it is worth mentioning that other solutions are possible, either following the sensor-based approaches described in the previous chapters, or using new controllers, as long as each of these controllers generate a cartesian velocity on a common frame.

7.3.1 Position Controller

The localization-based position controller is in charge of computing the required hand velocity for grasping and interaction, based on world-to-robot and world-to-object information provided by localization algorithms. Thus, we assume that the robot is able to obtain an initial estimation of the robot pose and target object pose in a world frame, either based on laser localization, sonar, an intelligent environment, etc. These initial estimations are denoted as ${}^W\mathbf{M}_R$ and ${}^W\mathbf{M}_O$, where W stands for the world frame, R is the robot base frame and and O is the object main frame, as shown in Figure 7.2. We assume that the physical interaction planner has already chosen a suitable task-oriented hand preshape, a hand frame, H, and a grasp frame, G, related respectively to the robot end-effector and to the object frame through ${}^E\mathbf{M}_H$ and ${}^O\mathbf{M}_G$.

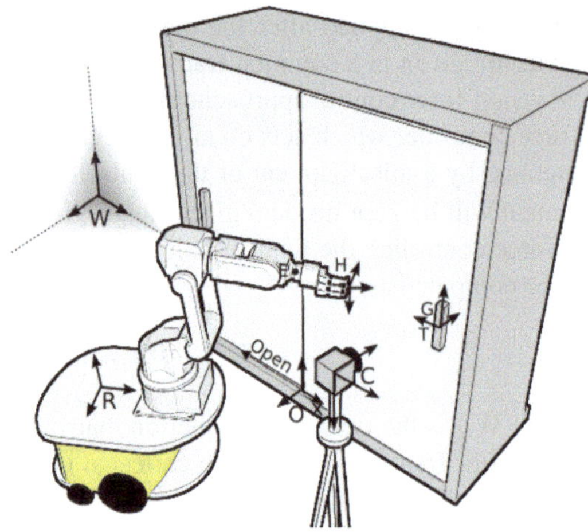

Fig. 7.2 We consider a mobile manipulator which has to perform a physical interaction task with an object. The mobile manipulator is located at frame R given with respect to the world frame, W. The camera is at frame C, and calibrated with respect to R. The object frame, O is supposed to be known in world coordinates. Finally, the physical interaction frames, H, G and T are computed by the physical interaction planner, and given with respect to R and O respectively.

First, the hand frame is transformed to world coordinates, as:

$$^{W}\mathbf{M}_H = {}^{W}\mathbf{M}_R \cdot {}^{R}\mathbf{M}_E \cdot {}^{E}\mathbf{M}_H \tag{7.3}$$

where $^{R}\mathbf{M}_E$ is the end-effector frame pose with respect to the robot base, which is assumed to be known from the robot kinematic model. The grasp frame is also transformed to world coordinates as:

$$^{W}\mathbf{M}_G = {}^{W}\mathbf{M}_O \cdot {}^{O}\mathbf{M}_G \tag{7.4}$$

Then, the hand-grasp relationship can be computed as:

$$^{H}\mathbf{M}_G = \left({}^{W}\mathbf{M}_H\right)^{-1} \cdot {}^{W}\mathbf{M}_G \tag{7.5}$$

and a proportional position-based control can be performed with the following equation, where λ_p is the control gain, and $\mathbf{h}^*_H$ is a pose vector build from the homogeneous matrix $^{H}\mathbf{M}^*_H$ (i.e. $^{H}\mathbf{M}^*_H = {}^{H}\mathbf{M}_G \cdot \left({}^{H}\mathbf{M}^*_G\right)^{-1}$):

$$\mathbf{v}^p_H = \lambda_p \mathbf{h}^*_H \tag{7.6}$$

This simple control law drives the hand in a straight line in order to reach the desired relative pose between the grasp frame and the localization-based estimated pose of the grasp frame on the target object. Unfortunately, robot localization errors are typically of a few centimeters [142]. Although these errors might be admissible for some applications, better accuracy is needed in most of the cases, as that provided by vision and tactile information.

7.3.2 Vision Controller

The object pose estimation provided by the VVS process described in chapter 6, is used here to compute a more accurate approximation of the grasp link matrix, as:

$$^{H}\mathbf{M}_G = \left(^{C}\mathbf{M}_H\right)^{-1} \cdot {}^{C}\mathbf{M}_O \cdot {}^{O}\mathbf{M}_G \tag{7.7}$$

where $^{C}\mathbf{M}_O$ is the object pose estimation computed by VVS, and $^{C}\mathbf{M}_H$ is assumed to be known, either because it is directly observed, as in the practical implementation of chapter 6, or because it is computed from camera external calibration and robot kinematics, i.e:

$$^{C}\mathbf{M}_H = {}^{C}\mathbf{M}_R \cdot {}^{R}\mathbf{M}_E \cdot {}^{E}\mathbf{M}_H \tag{7.8}$$

As full 3D information is available, a position-based visual servoing approach [113] can be adopted in order to obtain straight trajectories in cartesian space. The $^{H}\mathbf{M}_G$ matrix is used for setting the visual feature vector to $\mathbf{s}_v = \left(^{H}\mathbf{t}_G, \mathbf{u}\theta\right)^T$, where $^{H}\mathbf{t}_G$ is the translational part of the $^{H}\mathbf{M}_G$ homogeneous matrix, and $\mathbf{u}\theta$ is the axis/angle representation of the rotational part. Similarly, the desired feature vector $\mathbf{s}_v^*$ is set from the desired grasp link state, $^{H}\mathbf{M}_G^*$, as specified in the physical interaction task. The hand velocity computed by the position-based visual servoing control law is given by:

$$\mathbf{v}_H^v = -\lambda_v \widehat{\mathbf{L}_{s_v}^{+}} (\mathbf{s}_v - \mathbf{s}_v^*) \tag{7.9}$$

where the interaction matrix, $\widehat{\mathbf{L}_{s_v}^{+}}$, takes the form of expression 6.16.

This expression drives the robot hand in order to reach the desired reference of the grasp link matrix, $^{H}\mathbf{M}_G^*$. It allows for object reaching, but also for maintaining a suitable hand-object alignment during interaction. It is worth noting that the vision controller is independent of the method used for estimating the object pose. Even though the VVS method has been adopted in this particular case, another different method could be used as long as it is suitable for inclusion in a control loop.

7.3.3 Tactile Controller

We propose a tactile controller that looks for the alignment between the robot fingertips and a planar surface, such as a handle. Although it has been specifically designed for performing alignment tasks, it could be easily adapted for a different purpose, as long as a control velocity for the set of cartesian directions which can be robustly and accurately controlled with tactile information is provided. Depending on the sophistication of the tactile sensors, the sensor distribution, the hand configuration, and the task, more or less directions could be controlled.

We consider a Barrett Hand with one tactile array sensor on each fingertip, providing pressure distribution and magnitude information in a 8x15 pressure matrix,

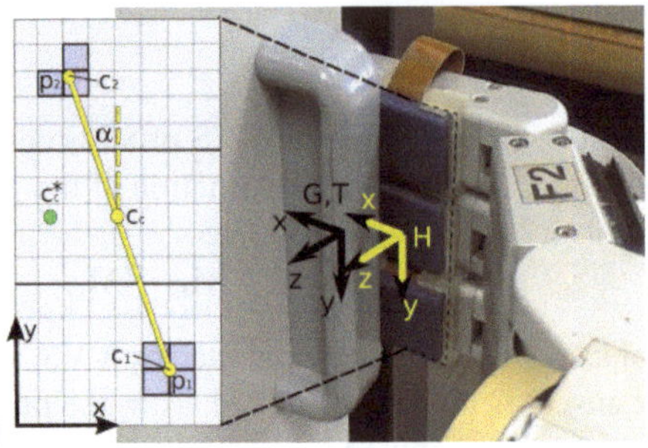

[t]

Fig. 7.3 The Barrett Hand in a hook precision preshape, with the tactile sensors installed at the fingertips. *H* is the hand frame, and *G* denotes the grasp frame. The biggest contact blob on sensors 1 and 2 is selected, and the centroids of these contacts are computed, together with the maximum pressure on each sensor and the angle between the contact line and the vertical.

as shown in Figure 7.3. First, the biggest contact blob on sensors 1 and 2 is selected and its centroid is computed, giving the points $\mathbf{c}_1 = (\mathbf{c}_{1_x}, \mathbf{c}_{1_y})$ and $\mathbf{c}_2 = (\mathbf{c}_{2_x}, \mathbf{c}_{2_y})$ in the sensor frame. The maximum pressure sensed on each of the two contact blobs are denoted as p_1 and p_2. The point $\mathbf{c}_c = (\mathbf{c}_{c_x}, \mathbf{c}_{c_y})$ is computed as the middle point between $\mathbf{c}_1$ and $\mathbf{c}_2$. Finally, α is computed as the angle between the line joining $\mathbf{c}_1$ and $\mathbf{c}_2$ and the vertical.

Three cartesian DOF's at the hand frame (*H* in Figure 7.3) are controlled in order to accomplish three goals:

- First, rotation around X axis is controlled in order to guarantee that the pressure is equally distributed between the tactile sensors, thus ensuring that all the tactile sensors keep the contact:
$$w_x = k_p (p_2 - p_1) \tag{7.10}$$
- Second, rotation around Z axis is also controlled in order to regulate α to zero. The goal is to be aligned with the handle.
$$w_z = k_\alpha \alpha \tag{7.11}$$
- Finally, translation along X axis is controlled in order to bring the point $\mathbf{c}_c$ towards a reference $\mathbf{c}_c^* = \left(\mathbf{c}_{c_x}^*, \mathbf{c}_{c_y}^*\right)$, which indicates the part of the tactile sensor where to keep the contact:
$$v_x = -k_c \left(\mathbf{c}_{c_x} - \mathbf{c}_{c_x}^*\right) \tag{7.12}$$

k_p, k_α and k_c are the control gains for each controlled direction. The velocity on the rest of directions is set to zero:

$$\mathbf{v}_H^t = (v_x, 0, 0, w_x, 0, w_z) \tag{7.13}$$

The selection matrix for the tactile controller is set to $\mathbf{S}_t = \mathbf{diag}(1,0,0,1,0,1)$ for this particular case. In the cases where tactile information is not available, such as in the phase of reaching, $\mathbf{S}_t$ can be set to zero so that the hand is controlled by position-vision-force integration.

In conclusion, when enough tactile information is available, tactile control can ensure that an accurate alignment between the hand and the handle is kept, by controlling just three cartesian DOF's. Unlike the position and vision controllers which are general and support many different types of hand-object alignments, the tactile controller proposed in this section is specific for alignment tasks with hook precision preshapes. This is due to the very local nature of the tactile sensors, which are installed only at the fingertips of the robotic hand. In the case of more advanced tactile sensors or different alignment tasks, it would be possible to control additional DOF's.

7.3.4 Force Controller

Finally, an active stiffness control with an explicit force reference is performed on top of the other controllers:

$$\mathbf{v}_H^f = \widehat{\mathbf{L}_\times}^{-1} \mathbf{K}^{-1} ({}^H\mathbf{D}_F \cdot \mathbf{f}_F - \mathbf{f}_H^*) \tag{7.14}$$

where $\mathbf{f}_F$ represents the force measured at each iteration at the force sensor frame, F, whereas $\mathbf{f}_H^*$ is the force reference used for pushing in the task direction, expressed in the hand frame. ${}^H\mathbf{D}_F$ represents the wrench transformation matrix between frames F and H (i.e. ${}^H\mathbf{D}_F = {}^H\mathbf{W}_F^T$) [89]. The force reference, $\mathbf{f}_H^*$, is computed from the task reference, $\mathbf{f}_T^*$ (i.e. $\mathbf{f}_H^* = {}^H\mathbf{D}_T\mathbf{f}_T^*$), which can be set to zero on those directions where a passive behavior is desired, but must take a value for those task directions which are explicitly force-controlled (those specified by the selection matrix $\mathbf{S_f}$).

7.4 Vision-Tactile-Force Suitability for Physical Interaction

Vision-tactile-force control provides a reliable approach to physical interaction, including all the advantages of vision, tactile and force sensors. The addition of tactile feedback allows to control some DOF's that could not be controlled with vision-force approaches. Tactile sensors provide a very accurate estimation of some components of the hand-grasp transformation, and allow to control them in a very accurate and reliable manner. In particular, non-constrained DOF's, which cannot be force-controlled, and for which vision cannot provide an accurate solution, can be taken into account with tactile control.

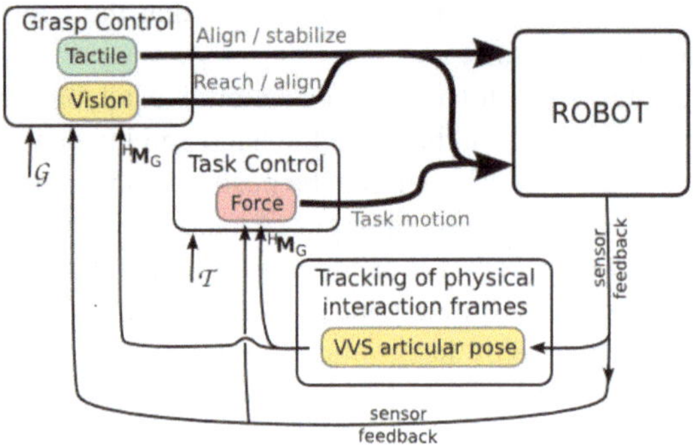

Fig. 7.4 The general vision-tactile-force control scheme in terms of the physical interaction framework

7.4.1 Vision-Tactile-Force Control for Grasping

Pick and place is one of the most common actions that a service robot needs to perform in home environments. A useful robot must be able to grasp an object from one place, and deliver it to the user, or either moving it to another place when tidying a room, for example. During the transport action, the object is always subject to the gravity force, and may be also affected by vibrations in the robot, or dynamic forces such as inertia.

For this reason, pick and place actions must be performed through stable grasps which are capable of resisting all the potential external disturbances. More concretely, it is desirable that contacts are performed on planar or concave surfaces, so that the contact region is as large as possible. Convex parts, edges and vertex normally generate unstable point or line contacts and should be avoided when possible.

Vision and force feedback is usually not enough for ensuring the stability of the grasp. Although vision can potentially provide the approximate position and orientation of the object, the accuracy of this information is not enough for reaching a desired contact position. In addition, during the grasping action, some object parts are normally occluded by the hand, which makes vision processing more difficult and introduces further inaccuracies. On the other hand, force sensors can, in some cases, detect contact with the object, but they do not provide any information regarding the contact location and type. Thus, neither vision nor force sensors are suitable for measuring the stability of a grasp.

Non-prehensile grasps can also benefit from the use of tactile feedback. Although grasp stability is not a crucial issue in this case, it is normally desirable to obtain a desired relative positioning between the robot hand and the object. Force sensors may provide incomplete information, specially on those directions which are not constrained. Vision can achieve a rough initial approximation which depends on the number and type of visual features. Tactile information can provide valuable additional information for establishing a precise and reliable contact.

Tactile sensors can normally measure interesting contact properties. In particular, tactile array sensors can provide a pressure matrix which can be used to recognize contact patterns, such as point, line or planar contacts. In addition, through the contact the robot establishes a direct relationship with the environment which allows to accurately locate the object. If the object model is available, an active sensing approach can be adopted in order to estimate precisely the object position and orientation, and then control the robot fingers towards the planned contacts. Therefore, tactile feedback can be used, either in an event-driven fashion or inside a control loop, in order to achieve a stable grasp, where all the fingers keep a suitable contact configuration.

7.4.2 Vision-Tactile-Force Control for Interaction

When a robot is performing compliant motion and following a constrained trajectory, a prehensile grasp provides rich force information that is normally enough for a correct execution without the need of other sensors, as it was shown in chapter 5. A prehensile grasp constrains all the relative hand-object degrees of freedom, and, therefore, any relative displacement between them generates a force through the contacts, which can be used inside a control loop in order to correct the misalignment.

However, an important number of our daily manipulation actions are performed with non-prehensile grasps, either because it is more appropriate for the particular task, or because it is not possible to apply a prehensile preshape. In this cases, any misalignment on a non-constrained direction can not be detected by the force sensor. For example, when opening a sliding door with a hook precision preshape, only the opening direction and the direction that points towards the door are position-constrained. Therefore, during the task execution, the force sensor can only provide information on these directions and not on the rest. In most of the cases, this partial information is not enough for successfully completing the task.

In the cases where force does not provide enough feedback about the task state, vision sensors can contribute with additional information. However, most vision processing algorithms are not precise enough and can often fail due to changing lighting conditions, occlusions, etc. During manipulation tasks, it is necessary to use precise and robust sensors, because a sensor failure may lead to catastrophic results.

Tactile sensors provide reliable and very precise information which can be used to complement force and vision feedback when their contribution is imprecise or incomplete. Tactile information can be used to detect hand-object misalignments and adapt the hand motion accordingly. In general, they are necessary for performing those tasks where the DOF's constrained by the grasp are not enough to ensure a successful execution only by force sensing.

7.5 Practical Implementation: The UJI Service Robot

In order to study the benefits that tactile feedback can provide to vision-force-based manipulation, a physical interaction task is performed with an articulated object when there is not enough information in the image features for an accurate vision-based 6D pose estimation, and where the relative hand-object motion is not completely constrained.

7.5.1 *Task Description*

We consider the task of opening a cabinet door of sliding type over 25 cm. The robot is manually moved in front of a cabinet as shown in Figure 7.2. The camera is placed in order to get a partial view of the cabinet door, and a coarse estimation of the homogeneous matrix describing the relationship between the camera frame and the robot base frame, ${}^{R}\mathbf{M}_{C}$, is previously calibrated by attaching a pattern to the robot hand and computing its pose with the Dementhon algorithm [38], and then making use of the robot kinematic model. Note that this step would not be necessary in a humanoid system, for example, where the eye-to-hand relationship can be approximately computed through robot kinematics. The door pose in the camera frame, ${}^{C}\mathbf{M}_{O}$, is coarsely calibrated in our case, although it could be also computed from robot laser and sonar-based localization algorithms.

We assume that the cabinet has been previously recognized, and that a structural model of the door is available for planning and vision-based tracking purposes. The physical interaction planner of chapter 4, automatically classifies the door handle with the *Fixed handle* label, and the *Fixed Handle Push* action is planned, leading to (see also Figs. 7.2 and 7.3):

- A task-oriented hand preshape suitable for performing the particular grasp and task. In this case, a hook precision preshape is planned.
- A hand frame, H, associated to the task-oriented hand preshape, attached to the part of the hand used for the grasp, and known with respect to the manipulator end-effector frame through hand kinematics. This homogeneous transformation is denoted by ${}^{E}\mathbf{M}_{H}$.
- A grasp frame, G, attached to the part of the object where the hand must be moved to, and expressed with respect to the object frame, O, through the homogeneous transformation matrix ${}^{O}\mathbf{M}_{G}$. ${}^{H}\mathbf{M}_{G}^{*}$ and $\mathbf{S_c}$ are set to the identity matrix so that a rigid hand-handle relative pose is desired.
- A task frame, T, placed at the same position that the grasp frame, so that the interaction task can be specified as a desired velocity along Z axis, $\mathbf{v}_T^* = \left(0,0,v_z^*,0,0,0\right)$. Therefore, $\mathbf{S_f} = \mathbf{0}$.

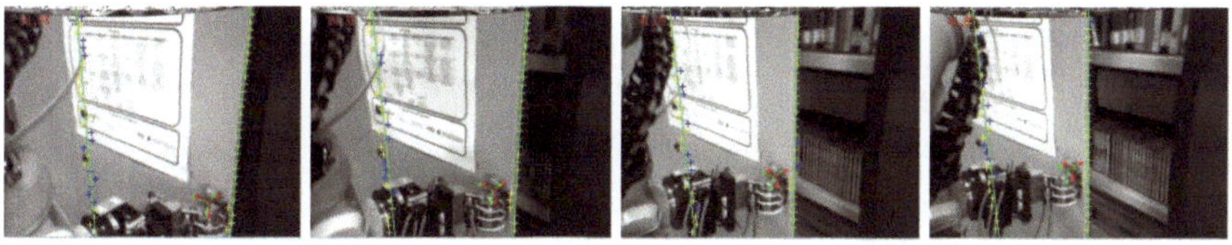

Fig. 7.5 The vision effects of introducing an error around Y axis in the pose of the grasp frame. VVS is able to track the articulated pose, but errors on the rest of directions cannot be corrected (note that the left edge estimation does not correspond to the real one).

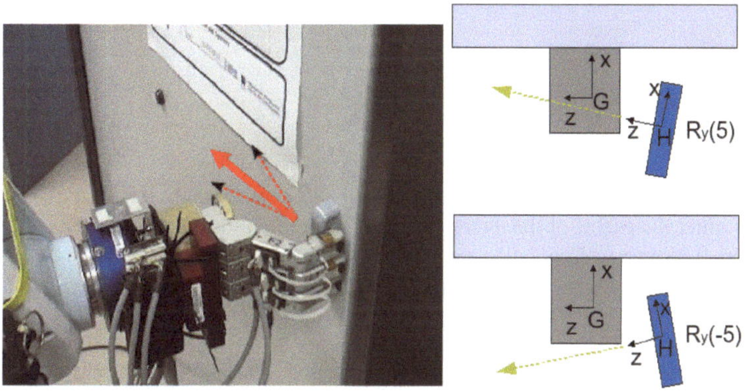

(a) (b)

Fig. 7.6 The task is to push open a sliding door under manually introduced errors in the initial estimation of the grasp link. For a rotational positive error around Y axis of the grasp frame, *G*, the hand motion has a positive component along the X axis of the real grasp frame, which finally leads to a frontal contact if not corrected. For a negative error, the robot pushes along a direction which has a negative component along the X axis of the real grasp frame.

The vision controller is based on the pose estimation given by the VVS method explained in section 6.2.2.2. Instead of using special markers like in the previous section, the natural object edges are tracked. However, only the left and right edges of the door are visible from the camera position, leading to a feature set which is not rich enough for getting a full rank interaction matrix. For this reason, the pose estimation algorithm is configured to track the cabinet door only along one DOF, corresponding to the opening direction, by setting $\mathbf{S}^{\perp} = \mathbf{diag}(1,0,0,0,0,0)$. Therefore, the pose estimation is unable to converge to the real pose on the rest of directions, and, thus, the vision controller cannot reliably control these directions.

Similarly, the hook precision non-prehensile grasp only finds position constraints along the frontal direction (*X* axis of the hand frame), and the opening direction (*Z* axis of the hand frame). As the rest of directions are not position-constrained, misalignments on these directions do not generate external forces, and, thus, cannot be detected and controlled with force feedback.

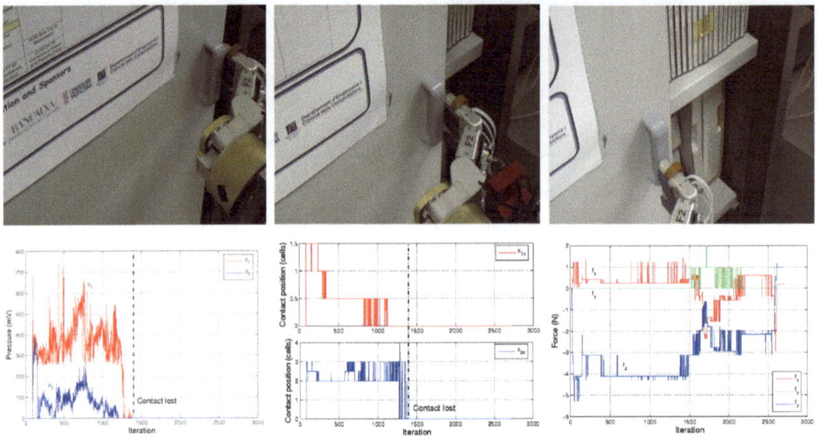

Fig. 7.7 Vision-force performance for an error of -5 degrees in Y axis of the handle pose. Top row: three snapshots of the interaction task, where it is shown how contact is lost during execution. Bottom row, from left to right: pressures at the fingertips (p_1 and p_2), X component of the contact centroids (c_{1_x} and c_{2_x}) and forces in the hand frame ($\mathbf{f}_H$).

Tactile sensors can take control of these directions when the hand is in contact. In order to study the benefits that tactile sensors provide, the door opening task has been reproduced several times, with three different sensor combinations and manually set rotational errors in the initial estimation of the object pose, simulating localization errors. Note that these errors are added to those already existing due to the poor calibration of the initial camera-robot and camera-object transformation.

In particular, an error of ± 5 degrees has been forced on each of the axis of the grasp frame, G, so that there is an important misalignment between the real orientation of the handle, and the one that the robot estimates (see Figure 7.6). For each error (positive and negative), on each axis, the task has been executed, first by using only the force sensor, then adding the vision modality, and finally by a combination of vision, force and tactile sensors. Therefore, a total of 18 trials have been performed, 6 for force-only, 6 for vision-force and 6 for vision-force-tactile, and a trial was considered as a failure when the robot was unable to open the door over 25 cm.

7.5.2 *Results*

Simple force control succeeded in only 3 out of 6 trials, whereas vision-force completed the task in 5 experiments, and vision-force-tactile performed well in all the 6 cases. In addition, the vision-force-tactile combination was the only one able to avoid undesired forces in directions other than the task direction.

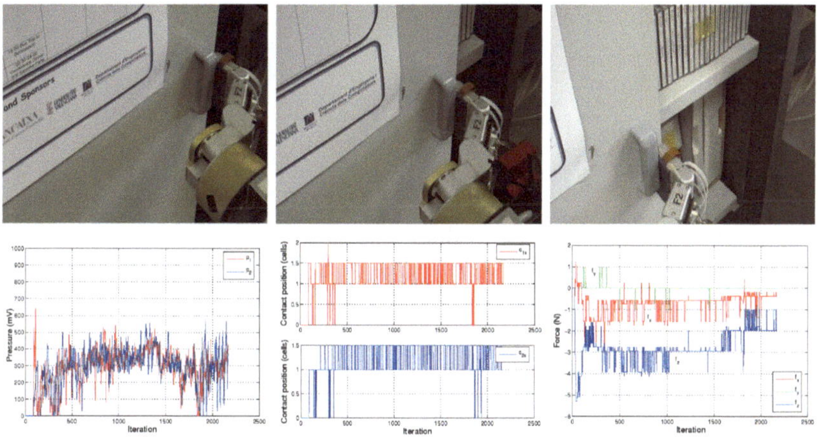

Fig. 7.8 Vision-tactile-force performance for an error of -5 degrees in Y axis of the handle pose. Pressure is balanced and contact is kept until the end of the task. Top row: three snapshots of the interaction task. Bottom row, from left to right: pressures at the fingertips (p_1 and p_2), X component of the contact centroids (c_{1_x} and c_{2_x}) and forces in the hand frame ($\mathbf{f}_H$).

Detailed results for the interesting case of a rotation error around Y axis are shown in Figures 7.7, 7.8 and 7.9. In the case of Figures 7.7 and 7.8, the introduced error is manifested in a misalignment that makes the robot push along a direction which has a negative component along the X axis of the real grasp frame, G, as shown in Figure 7.6b. As this direction is not position-constrained, force is not able to detect the misalignment. Similarly, the vision part has partial information and can only track the task DOF. Thus, the articulated pose estimation still contains the initialization error. An opening strategy using only vision and force sensors would easily lose contact, as shown in Figure 7.7. However, the vision-tactile-force approach is able to perceive contact information, and controls the robot so that a contact is always present at the desired location in the fingertip (Figure 7.8). The force disturbances in the case of vision-force are due to a frontal contact of the finger with the handle, appearing at the moment of losing the contact, which generates a small frontal force.

It is also worth noting that vision-tactile-force is able to balance the pressure between the fingertips, whereas vision-force is not able to detect and control this issue. This is clearly shown in Figure 7.9, which illustrates the case of a positive rotation error around Y axis in the localization of frame G. In this case, the pushing direction has a small positive component in X axis, which slowly drives the fingertip towards the door, as shown in Figure 7.6. If only vision sensors were considered, frontal collision could not be detected, causing important damage to the robot. Figure 7.9 shows the behavior of vision-force and vision-force-tactile control in this case. Note that, under vision-force, there is contact only with one fingertip since the very beginning (7.9.a-b), and vision-force is not able to correct this misalignment. Consequently, the whole task force is made by only one finger, which has to support

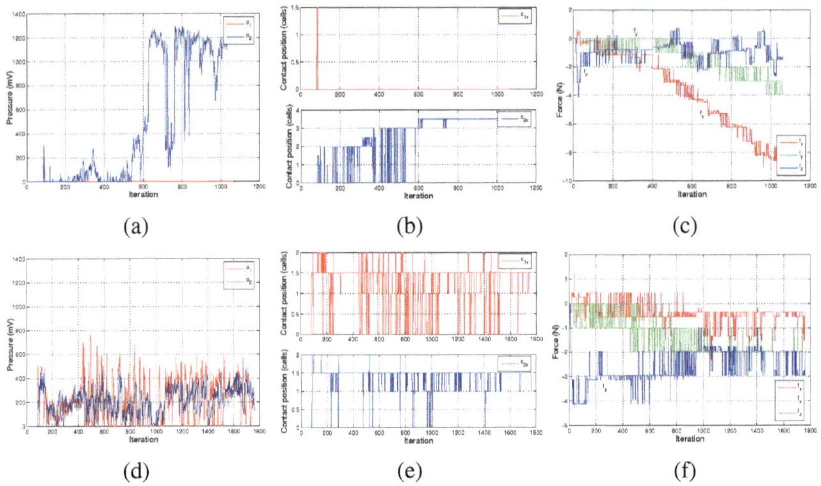

Fig. 7.9 Results for an error of +5 degrees in Y axis of the handle pose. Vision-force control is not able to completely align the hand. Vision-force-tactile aligns the hand successfully and distributes the pressure between the fingertips. Top row: vision-force; Bottom row: vision-tactile-force. From left to right column: pressures at the fingertips (p_1 and p_2), X component of the contact centroids (c_{1_x} and c_{2_x}) and forces in the hand frame ($\mathbf{f}_H$).

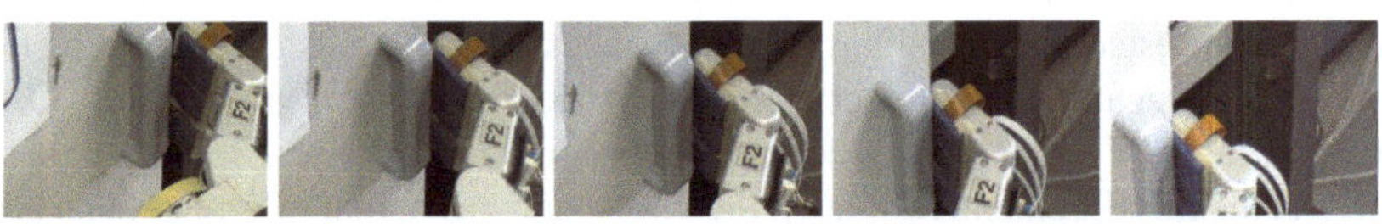

Fig. 7.10 A sequence showing the vision-tactile-force alignment process during task execution, starting from a coarse initial position

a high pressure, increasing the risk of sensor or mechanics damage. Vision-force-tactile, however, is able to balance the pressure, ensuring contact with all the fingers (Figure 7.9.d-e). Even if vision-force is finally able to complete the task, note that, as a consequence of the initial introduced error, the fingertip finally makes frontal contact with the door, leading to a high force in the frontal direction that exists almost from the beginning of the execution (7.9.c). As expected, vision-force-tactile avoids this situation, keeping the pressure level, contact position and forces inside a normal range.

Figure 7.10 shows a sequence of the vision-tactile-force execution, whereas Figure 7.11 shows the corresponding *tactile image*. It can be appreciated how, starting from an initial position with important alignment errors, vision-tactile-force integration is able to correct them and converge to a reliable and safe configuration where all the fingertips are in contact and aligned with the handle.

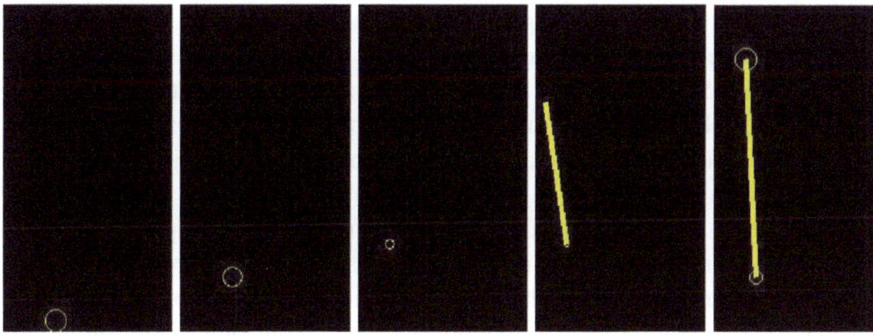

Fig. 7.11 A visual representation of the evolution of the contacts during the alignment task. A circle represents a point contact. The radius of the circle indicates the pressure magnitude. Two point contacts generate a line contact.

7.6 Discussion and Conclusions

We have presented a new approach for combining vision, tactile and force feedback into a global control law. This approach is based on the concept of hybrid control: for a given cartesian degree of freedom, the most confident available sensor information between tactile, vision, and position sensors is selected. Force information, however, is used on all the degrees of freedom in order to ensure a safe physical interaction.

We propose to give priority to tactile sensors over vision and position information. However, it could be different in other cases, like in [146], for example, where localization algorithms provide very accurate pose information. In order to select which sensor controls each DOF, the corresponding selection matrices $\mathbf{S}_p$, $\mathbf{S}_v$ and $\mathbf{S}_t$ can be set accordingly. It is worth noting the possibility to modify the sensor assignation at run time. If, for example, vision processing fails at a given moment, it would be possible to remove the DOF's assigned to the vision controller, and assign them to the position controller, or tactile controller in case they have enough information to control them. Similarly, position-vision-force control can be performed when tactile information is not available, as in the phase of reaching, for example. This could be managed automatically following, for example, the *sensor resolvability* approach [131], that allows to identify which sensor provides the most suitable estimation for controlling a particular cartesian direction.

Due to the limitations of our tactile sensors, and the kind of tasks considered, only three DOF's can be accurately controlled by the proposed tactile controller. The rest is controlled by vision, or by the position controller in case vision is not available. However, the particular tactile, vision, position and force controllers are independent of the global scheme, in the sense that it would be possible to design a new tactile controller able to control six DOF's in the case of using tactile sensors which provide enough information.

Regarding the vision controller, it would be desirable to control all the six cartesian DOF's, or, at least, those which are not already controlled by the tactile controller. We have discussed in chapter 6 the difficulties to estimate a full 6D pose when the robot is close to the object. In these cases, only partial views are usually obtained, the interaction matrix can take a very low rank, and local minima can appear in the VVS minimization process. We cannot assume that a rich feature set is always available. Therefore, errors in the visual estimation can always appear, making it necessary to use force and tactile sensors to deal with such errors during physical interaction.

In summary, a vision-tactile-force control approach has been proposed and validated in a real environment. A door opening task has been executed through the combination of the control signals provided by a position controller, which has an initial coarse estimation of the object pose, a vision controller based on the articulated object pose computed by the Virtual Visual Servoing method, a tactile controller, which looks for not losing the contact during manipulation, and a stiffness force controller, in charge of pushing along the task direction at the same time that the force is regulated on the rest of directions. Different sensor combinations, such as force-only, vision-force or tactile-force, are also possible in case that one or more sensors become unavailable. In order to study the advantages of adding tactile feedback to vision and force-based manipulation, several experiments have been carried out with the task of opening a sliding door under manually introduced errors. Results have shown how vision-tactile-force integration is able to correct the hand-object misalignments generated by an inaccurate vision-based reaching, keeping a suitable grasp link during the whole task, and increasing the global reliability. Therefore, we conclude that introducing tactile feedback into the existing vision-force manipulation approaches can greatly improve robot physical interaction capabilities.

This chapter finishes with the set of three chapters devoted to the sensor-based execution of physical interaction tasks. Each of them has introduced a new type of sensor on top of the previous developments: force feedback in chapter 5, vision-force in chapter 6 and vision-force-tactile in this chapter. Although the use of force feedback exclusively already represents a robust solution, the addition of vision and tactile information allows to deal with those tasks for which force information might be incomplete, thus providing further reliability. The next chapter concerns the integration of all these concepts into a suitable software architecture, and describes the first steps towards the long term goal of building a multipurpose assistive robot.

Chapter 8
Towards an Assistive Mobile Manipulator

8.1 Brief Introduction

Service robotics is an area that has been receiving increased attention in the last years. However, there are very few service robots able to perform manipulation tasks in human environments, and even fewer which can perform more than one different action. Robots are usually programmed to perform a given, specialized task, and are unable to extend their capabilities to other domains. This chapter describes the first steps towards the long-term goal of building a multipurpose assistive mobile manipulator from the methods described in this book.

Building a multipurpose service robot is a challenging goal which requires the integration of versatile physical interaction planners with reliable low-level sensor-based controllers. Moreover, this integration must be done on the basis of a suitable software architecture providing appropriate sensor-control communication facilities, and allowing to switch between different controllers in real-time, according to the state of the physical interaction task. Throughout this book, we have proposed several methods for the specification, planning and robust control of physical interaction tasks. In this chapter, we propose a software architecture that integrates all these aspects, leading to a unified robotic system. The architecture is then implemented on a real robotic platform, with the long-term goal of offering a great variety of physical interaction capabilities in a multipurpose service robot.

More concretely, we describe the UJI Service Robot, which serves as a platform where to apply the different contributions of our work. The robotic system, which is described on section 8.2 of this chapter, provides an integrated platform for defining and executing daily manipulation tasks in a modular and flexible way. The sensor-based control methods described in chapters 5, 6 and 7, provide reliable solutions to the execution of physical interaction tasks specified with the framework and planner of chapters 3 and 4. In section 8.3 we propose a software architecture that integrates all the previous developments. It allows to switch between the different states of the physical interaction task according to sensor-based conditions, thus implementing an event-driven approach on top of the low-level controllers.

M. Prats et al.: Robot Physical Interaction through the Combination of Vision, STAR 84, pp. 129–157.
springerlink.com

The UJI Service robot has been endowed with two main physical interaction capabilities: opening doors and grasping books. These capabilities are described in detail in section 8.4. The book grasping capability represent the basis of the more specialized application of the UJI Librarian Robot, described in section 8.5, designed to look for a book in a library and deliver it to the user who requested it.

However, the framework for physical interaction, together with the physical interaction planner, offer a great number of useful manipulation capabilities, still to be explored. In section 8.6, we make an overview of some useful physical interaction tasks, and show how they are supported by the physical interaction framework and planner. In order to illustrate the results of the planner, we apply the vision-based pose estimation methods presented in chapter 6 to real images, following an augmented reality approach. Although many aspects have still to be considered, the number of different situations supported by the planner envisions promising results in future versions of the UJI Service Robot.

8.2 Description of the System

The UJI Service Robot is a mobile manipulator consisting of a Mitsubishi PA-10 7 DOF's arm mounted on an ActivMedia PowerBot mobile robot, as shown in Figure 8.1. The manipulator is endowed with a three-fingered Barrett Hand and

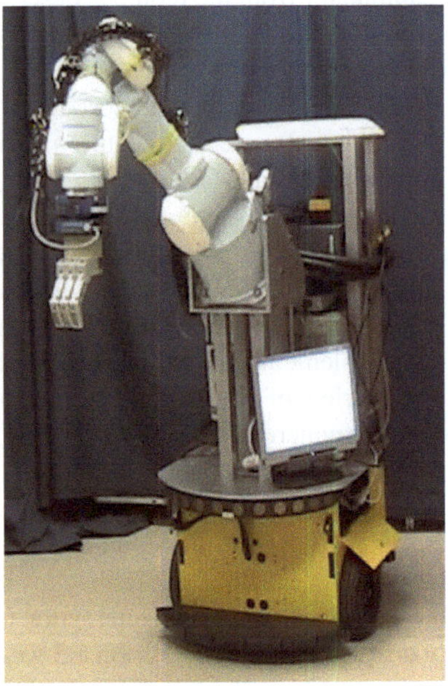

Fig. 8.1 The UJI Service Robot

a JR3 force, torque and acceleration sensor mounted at the wrist, between the hand and the end-effector. This sensor not only provides a six-dimensional vector of force/torque data, but also measures the linear and angular acceleration on all the axis. This additional data is useful to distinguish the inertial force produced by the movement of the arm itself from the force derived by collisions of the hand during the motion. A small IEEE1394 camera is also attached to the wrist of the robot. The manipulator, hand and camera controllers, along with the control computer, are mounted on the mobile platform and connected to the PowerBot batteries. The computer is a Pentium 4 running at 3 GHz, with 512 Mb of RAM.

The hand has been improved by adding arrays of pressure sensors on the finger-tips. The sensors consist of three 8 x 5 cell matrices, that cover the inner parts of the distal phalanxes of each of the three fingers. Each cell is a square of 2.3 mm side. The sensor principle is resistive and allows to detect a complete two dimensional pressure profile by the use of an homogeneous sensor material which is contacted by an adequate electrode matrix [188].

8.3 Control Architecture

Several software architectures for mobile robots have been proposed in the literature [57]. However, there are few contributions in the area of manipulators [93, 144]. As it is pointed out in [171], a control architecture for manipulators should be endowed with:

- The manipulation ability, in charge of acting on the environment.
- The sensory ability, capable of obtaining information from the world.
- The data processing ability, which processes data of system activity.
- The intelligent behavior ability, capable of modifying system behavior according to external information.

Therefore, a software architecture for manipulation must contain sensor information processing modules feeding low-level manipulation controllers. But an event-driven layer on top of these controllers is also needed, providing a “behavior” to the robot. The behavior layer is in charge of selecting the most appropriate low-level controllers according to the state of the task. Most of the behavior-based architectures implement a set of preprogrammed behaviors for a particular task [5]. Robots running under these architectures lack the capability of building new behaviors or modifying the existing ones during the ongoing execution.

From our point of view, this is a fundamental issue for future robot development. A robot must be able to execute the tasks it has been designed for, but also to modify or even to create new behaviors according to its experience, within a learning framework. Taking these goals into account, our architecture is structured into three main modules: perceptions, abilities and actions, as shown in Figure 8.2.

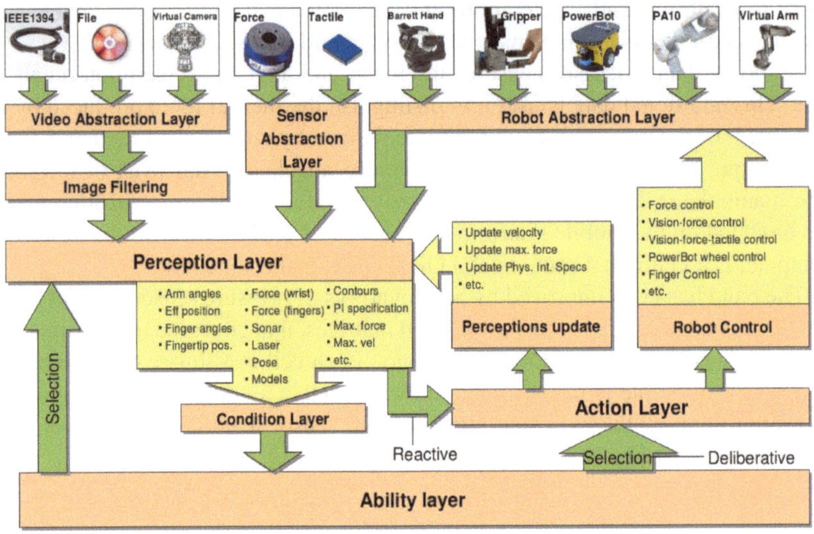

Fig. 8.2 The Architecture for Mobile Manipulation (AMM)

8.3.1 Perceptions

As perception, we understand any information that could be used to guide the robot actions. We consider both external and internal perception. External perception includes, for example, the perception of the force at the wrist or the robot's perception of the objects pose. Internal perceptions include proprioceptive information and the references, conditions and immediate goals of the robot. Some examples are the joint values, the end-effector position, the maximum force that the robot's fingers can support, the condition that compares the current force at the fingers with respect to the maximum allowed, the position in space where the robot has to immediately move its hand, or the desired position of the physical interaction frames.

The robot is also aware of *conditions* on most of these perceptions, as for example, whether the force at the fingers is greater than a given value, whether the task frame has reached a target reference, etc. Many of the perceptions depend upon others. For example, the perception that senses the force at the wrist makes use of the perception that stores the current effector position, in order to compute and to balance the gravity effects that the hand mass introduces into the force sensor. The architecture provides very simple mechanisms for defining these kinds of relationships, and for easily integrating new sensors. Hardware abstraction, extensibility and scalability are important features for any architecture [139].

In particular, the most important perceptions that the architecture implements are the following:

Internal perceptions: arm and hand joint values, the end-effector position with respect to a world frame, the hand frame, the distance of the hand frame w.r.t

another given frame, the time, the maximum force that the robot can resist, the physical interaction task specification, etc.
External perceptions: images, contours, objects pose, the force at the wrist, the force at each fingertip, pressure profile, etc.

8.3.2 Actions

The architecture distinguishes between three different types of actions:

Control actions. Data coming from perceptions are used as feedback to perform control actions in a reactive manner [5]. A control action takes as input a subset of perceptions and generates a control vector, which is sent to the robot. The sensor-based controllers presented in chapters 5, 6 and 7, have been implemented as control action modules in this architecture, along with other control algorithms that manage the Barrett Hand and the mobile robot wheels. As an example, the action that manages the force-based compliant motion of the arm, would take as input the physical interaction task specification, the current state of the physical interaction frames, the current force at the wrist, and internal perception such as joint values, the end-effector position, etc.
Perceptual actions. The architecture also allows for the execution of actions which do not control robot motion. Instead, these actions can act on the robot perception, as for example, modifying a task reference, the physical interaction frames, etc. As an example, the physical interaction planner of chapter 4 has been integrated into this architecture as an action which modifies the perceptual state of the robot. When the robot has to perform a physical interaction task, it executes the planning action, which automatically updates the required internal perceptions.
Behavioral actions. A third type of action allows to build robot *ablities* (explained in the next section) during execution. These actions are in charge of updating the robot behavior states. If, for example, a physical interaction task fails during execution, a behavioral action can be executed in order add the suitable recovery perceptual and control actions to the robot execution plan. Behavioral actions can be used to execute high-level task planning algorithms.

Control actions implement a direct connection between perception and control, providing the robot with a reactive behavior. Information coming from perceptual modules does not need to visit higher cognition modules before being used for action. Perceptual changes, including modifications in the physical interaction frames or maximum allowed velocity, for example, will be immediately reflected in the robot actions. Perceptual and behavioral actions allow to modify the internal state of the robot, either at the perception level, or at the high-level behavioral layer.

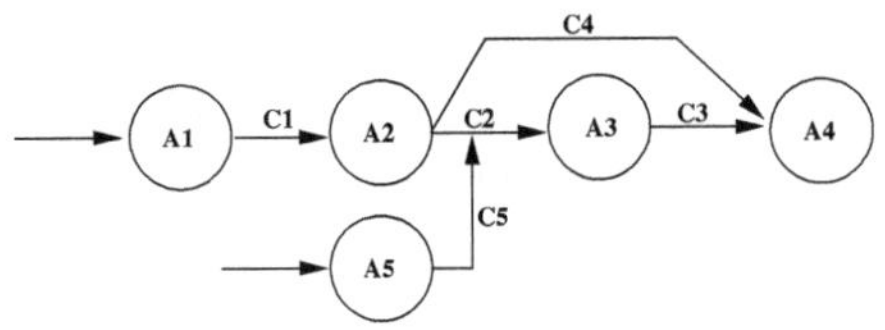

Fig. 8.3 An ability is the connection of actions by means of conditions

8.3.3 Abilities

Perception is used for guiding the robot actions at the lowest level, but also for making an internal representation of the world and planning future actions. Thus, perceptual information must also flow to high level modules where plans are made and supervised.

The *abilities* module is in charge of continuously selecting which perceptions to activate and which actions to perform according to these perceptions. It is in charge of selecting the robot behavior and supervising its performance. In terms of this architecture, we define an ability as a set of actions or other abilities connected by conditions. An ability can be seen as an automaton, where nodes are actions or other abilities, and arcs are conditions, which are built on perceptions.

The main difference with respect to other existing approaches, such as the skill primitive nets [93, 182, 67], is that abilities are not only composed of manipulation primitives, but also contain nodes that act on the robot internal perception, and even on the ability itself. The task information is not encoded within the action code. Instead, it is stored in perceptual modules that are independent of the action. This feature allows to have few and general motor actions, that can be used in many physical interaction tasks, after performing the suitable perceptual actions.

Figure 8.3 depicts an example of an ability that shows all the possible connections that the architecture supports. The nodes represent actions, whereas the arcs represent conditions. The architecture allows the parallel execution of actions. In the example of the figure, there are two initial nodes, A1 and A5. Thus, these two actions will execute concurrently. If condition C1 occurs when executing A1, the robot will start to execute A2 while still performing A5. If at this point condition C2 holds true, action A2 will finish but only action A5 will be active, because C2 will be waiting for C5 to trigger. If this happens, A3 will start its execution. If, instead, the condition that triggers when doing A2 is C4, then A4 will start its execution immediately, without waiting for C5 or C3. So, apart from the connections between actions, the architecture also provides facilities for synchronization.

Moreover, due to the recursive definition of *ability*, as the connection between actions or other abilities, it is allowed to integrate existing abilities into others. With this mechanism it is possible to construct more complex abilities incrementally. For example, a robot could know a simple ability such as moving along a direction until a high force is detected. This ability could be integrated into a more complex one such as pushing a button, which in turn could belong to another one, and so on.

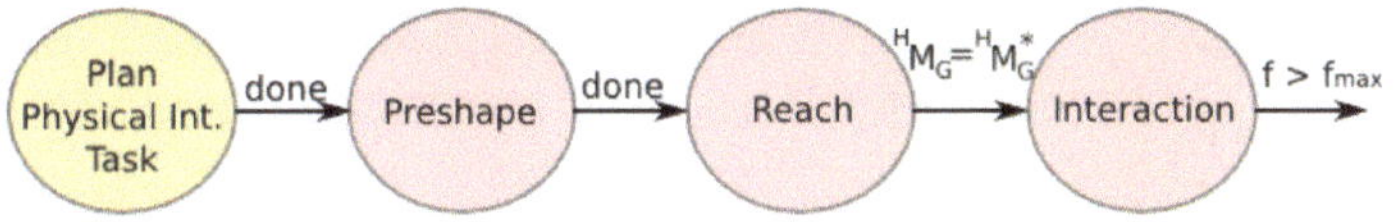

Fig. 8.4 Ability for turning a door handle

This offers a framework for robot learning that we would like to address in future work: the robot could incrementally learn new abilities on the basis of the skills that it already knows.

8.4 Physical Interaction Abilities

Using the software architecture described in the previous section, the physical interaction planner and the different sensor-based control schemes proposed in the previous chapters, door opening and book grasping abilities have been developed in the UJI Service Robot, as described in the following sections.

8.4.1 Door Opening

Different door opening abilities have been implemented using several sensor combinations: force-only, vision-force and vision-force-tactile. The first case can be normally applied to the task of opening ordinary doors, where the type of the grasp introduces enough cartesian constraints so that rich force information is available. The second and third cases allow to perform opening tasks where only a few directions are position-constrained, such as pushing open sliding doors.

Opening an ordinary door requires two consecutive physical interaction tasks: turning the door handle, and pulling open the door. The latter has been already studied as part of the practical implementations of chapters 5 and 6. In addition, the case of sliding doors has also been addressed in chapter 7. Therefore, we focus here on the task of turning the door handle.

Figure 8.4 shows a representation of the force based turn handle ability. It starts with a perceptual action which internally executes the physical interaction planner. It takes as input an object structural model and a high-level task description, and generates the specification of the physical interaction task, which is later executed through control actions implementing sensor-based controllers. Concretely, three control actions are defined: the preshaping action, in charge of controlling the hand to the desired hand posture; the grasping action, which performs the reach to grasp phase; and the interaction action, which actually performs the particular task motion. A force-based condition is added in order to detect the end of the task: when the opposite force is greater than a threshold value, the handle is considered to be completely turned.

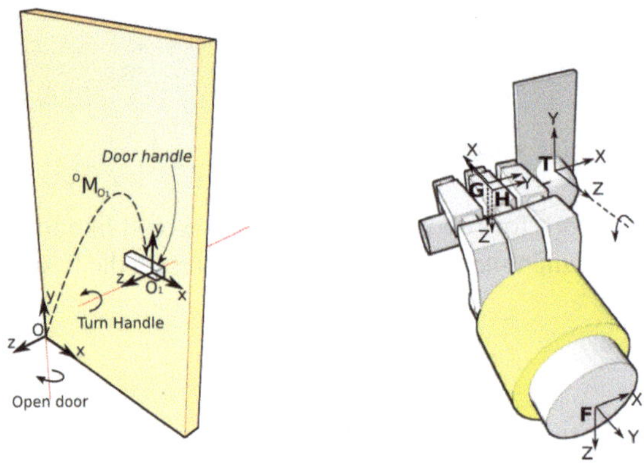

(a) A door structural model. (b) Frames during the task execution.

Fig. 8.5 Specification of the turn handle task in terms of the physical interaction framework

The structural model of the objects is implemented as an internal perception in the software architecture, together with a high-level task description in terms of *object tasks* and *object actions* (see section 4.4). The specification of the physical interaction task is also part of the robot perceptual state, including the physical interaction frames, the task-oriented hand preshape, velocity-force references, and the grasp link, which is computed from vision, tactile or force sensor readings, according to the case.

A more detailed description of this particular physical interaction task and its control approach is given below.

8.4.1.1 Planning the Physical Interaction Task

Figure 8.5 illustrates the specification of the handle turning task, in terms of the physical interaction framework, for a particular door. The structural model of the door, which is given as input to the physical interaction planner, contains a part which is labelled with the *Door handle* object class. The task is described at a high level as the *Door handle turn* object action. The physical interaction planner selects a hook power hand preshape, according to the guidelines given in section 4.5, and sets the hand frame and grasp frame accordingly, as shown in Figure 8.5b. The task frame is set to the handle revolute joint, so that the task is described in terms of the natural constraints as a rotation velocity around its Z axis. An explicit force reference is not needed in this task, thus $\mathbf{S_f} = \mathbf{diag}(0,0,0,0,0,0)$.

Fig. 8.6 A sequence of the door opening task

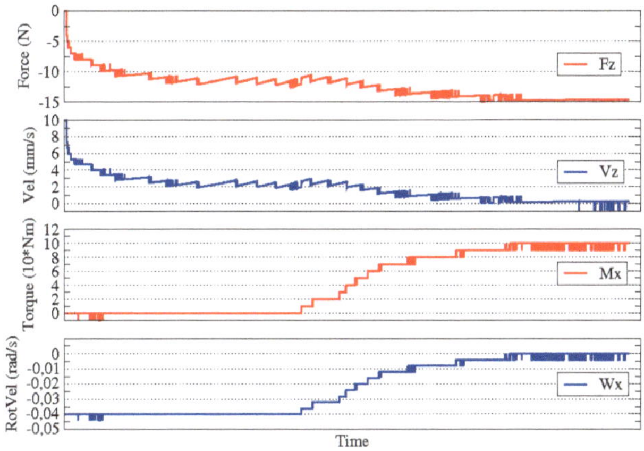

Fig. 8.7 Forces during the door opening task, and the corresponding velocity computed by the velocity-force control law

8.4.1.2 Task Execution

The interaction task is performed by means of a stiffness force controller which modifies the position/velocity signal, as it has been already described in the different practical implementations of the previous chapters. The task velocity, $\mathbf{v}_T^*$, is transformed from the task frame to the end-effector frame, following the kinematic chain defined by the physical interaction frames. The implicit force control on all the directions makes the hand adapt naturally to the constrained trajectory, and automatically stops motion in the task direction when the handle reaches the limit.

A sequence of the task execution is shown in Figure 8.6, whereas Figure 8.7 depicts the forces that appear during the task, $\mathbf{f}_F$, given in the force sensor frame. First, the absolute value of the force in Z direction of the force sensor frame increases until an approximate value of 10N which corresponds to the resistance of the particular door handle. During this stage there is no torque around Y direction, which means that the estimation of the task frame according to the model is good enough. Then, the handle starts offering more resistance, increasing the opposite force. At this point, the velocity, regulated by force control, decreases to zero, which corresponds to the case where the robot is applying the maximum possible force, and the task is considered as finished.

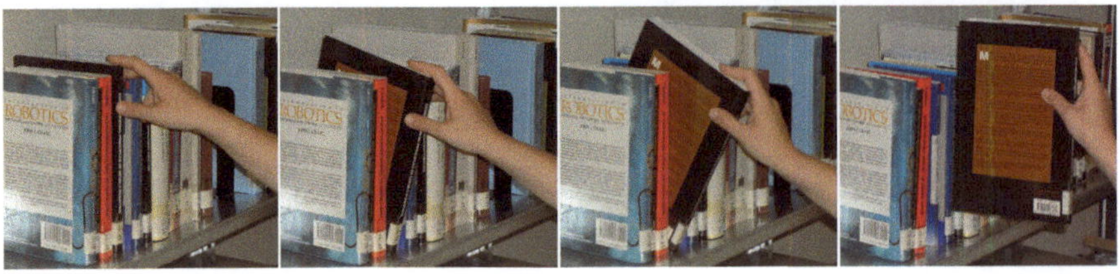

Fig. 8.8 Book extraction task performed by a human

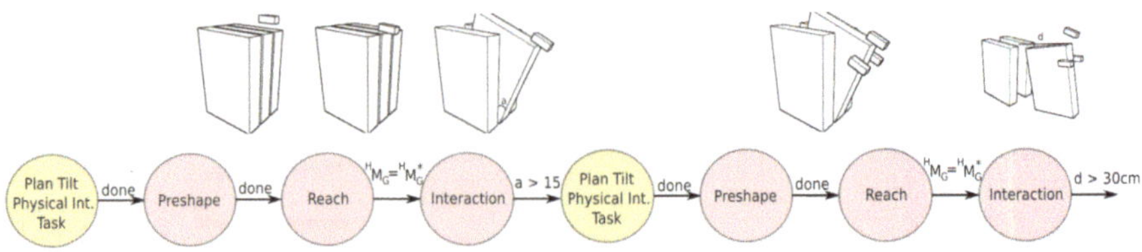

Fig. 8.9 A representation of the book grasping ability

8.4.2 Book Grasping

The goal of the book grasping ability is to extract a book from a shelf, while standing among other books. The approach is to do it as humans do: one finger is placed on the top of the target book and used to make contact and pull back the book, making it tilt with respect to the base, as shown in Figure 8.8. The book grasping process can be decomposed into two different physical interaction tasks: one in charge of tilting the book, and another one in charge of the extraction part.

Figure 8.9 shows an illustration of the book grasping ability. As in the previous case, it is composed of the planning action and the control actions. First, the physical interaction planner builds the specification of the first physical interaction task, thus setting the corresponding internal perceptions. Next, the preshaping, reaching and interaction control actions perform the tilting task. Finally, the extraction task is specified and executed through the corresponding control actions.

Both the tilting and extraction tasks have been implemented as a special case in the physical interaction planner. They are specified as follows:

8.4.2.1 Book Tilting Task

Planning the Physical Interaction Task

In Figure 8.10, the specification of the book tilting task, including the physical interaction frames, is shown. There are two possibilities for the task frame in this case. The first is to set it to the book base (frame T' in Figure 8.10), so that the task is described as a rotation velocity around this frame. The second possibility is to set the task frame on the top edge of the book (frame T in Figure 8.10), so that the task

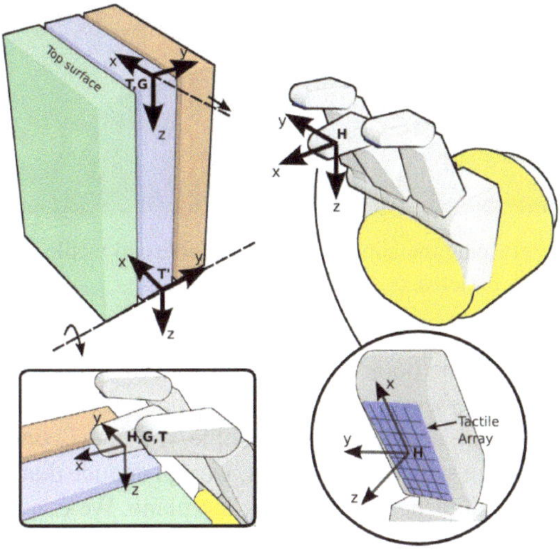

Fig. 8.10 The physical interaction frames involved in the book grasping task. The tactile array is used to estimate the relationship between the hand and the grasp frame, ${}^{H}\mathbf{M}_{G}$.

is described as a negative translational velocity along X direction. We have opted for the second solution, because it avoids the use of the book model. In the first case, the height of the book should be known in order to transform the task velocity from the task frame to the hand frame. By adopting the second solution, we make the approach general for any book size. Two references are set in the task frame, $\mathbf{v}_{T}^{*}$ and $\mathbf{f}_{T}^{*}$. The first one is set to a negative velocity in X axis, in order to perform the task motion, whereas $\mathbf{f}_{T}^{*}$ is set to a force along Z axis. This force is needed in order to make enough pressure on the book surface and avoid slip. Therefore, $\mathbf{S_f}$ is set to $\mathbf{diag}(0,0,1,0,0,0)$.

The one-finger precision hand posture is selected, where one of the fingers is slightly more closed than the other ones, so that the hand can easily make contact on the top of the book with one finger, as shown in Figure 8.10. The hand frame is set to the inner part of the middle finger fingertip, just in the center of the tactile sensor. The fingertip has to make contact on the top of the book. Therefore, the grasp frame is set to the book top surface. The desired relationship between the hand and the grasp frame, ${}^{H}\mathbf{M}_{G}^{*}$, is set to the identity.

Task Execution

This task is performed by means of force-tactile control: tactile information is used as feedback to a grasp controller in order to continuously improve the contact between the hand and the book, whereas force feedback is used to perform the task motion, coping with uncertainties and ensuring that a suitable force is performed on the book surface so that there is no slip.

Fig. 8.11 The robot grasping the book by means of force and tactile-based continuous estimation of hand-to-object relative pose

Contact on the top of the book is performed with the tactile array. A qualitative relative pose between the sensor surface and the book can be estimated according to the sensor cells that are activated. For example, if there is contact with the upper part of the sensor, but not with the lower part, we can deduce that the sensor plane is tilted around Y axis with respect to the book top plane. We consider that the finger is completely aligned with the book surface when there are cells activated on each of the four XY quadrants of the hand frame, i.e., all the tactile sensor surface is in contact.

Taking as input this qualitative description of the relative pose, a tactile-based grasp controller regulates the rotation around X and Y axis of the hand frame in order to achieve contact on each of the XY quadrants of the hand frame. With this approach, the behavior of the robot is completely reactive to the tactile sensor readings. The goal is to keep the sensor plane always parallel to the book top plane, thus ensuring that ${}^{H}\mathbf{M}_G = \mathbf{I}_{4\times4}$.

According to the task description, the task motion is performed by moving the hand along negative X axis of the task frame, while applying a force along Z axis, i.e. $\mathbf{f}_T^* = \left(0,0,f_z^*,0,0,0\right)$ and $\mathbf{v}_T^* = (-v_x^*,0,0,0,0,0)$. This motion makes the book tilt with respect to the base, as shown in Figure 8.11. Note that, as the fingertip moves backwards and the book is tilted, the tactile sensor may lose contact with the lower part. This situation is detected by the qualitative pose estimator, and corrected by the grasp controller, so that the hand frame is always aligned with the grasp frame, ensuring that task motion can successfully be transformed to end-effector coordinates.

8.4.2.2 Extraction Task

For the extraction task, the task frame is kept on the top surface of the book, but its rotation is updated in order to be aligned with the horizontal plane. A special preshape is adopted, called *three-finger spherical* preshape, where the hand frame is kept at the inner part of the middle finger fingertip, as in the previous case. This preshape allows to grasp the book using two fingers, whereas the third one remains in contact with the top of the book. A translational velocity reference is set along the extraction direction, and the interaction task is performed with simple velocity-force control.

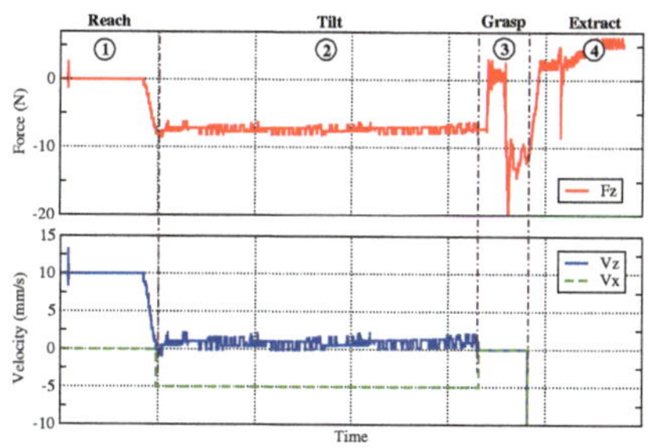

(a) Force and pressure during tilting.

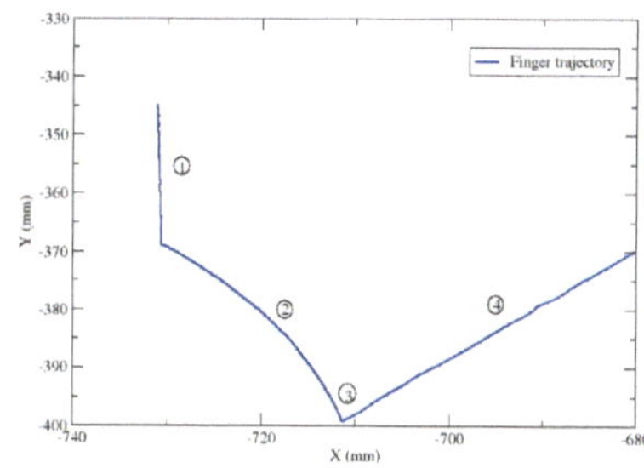

(b) Fingertip trajectory, represented in a world frame aligned with the floor.

Fig. 8.12 Results of the book grasping task

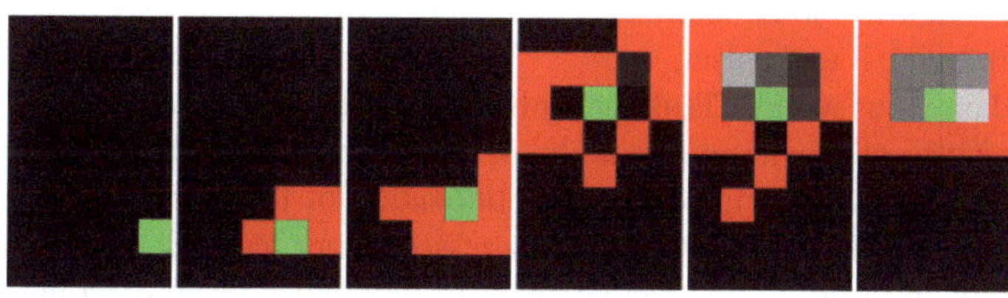

Fig. 8.13 Evolution of the tactile pattern during the book tilting task. Red pixels represent the boundary of the contact area, whereas the green point is the contact centroid.

Figure 8.12a represents the forces that appear during the whole task, and the consequences on the robot's behavior. The first stage corresponds to the initial reaching movement, when the fingertip is moving along the Z direction of the hand frame, searching for the contact. Thus, no forces are present, and the robot moves with a constant velocity along Z axis. When contact is made, the opposite force suddenly increases, until it reaches a reference value of 8N. At this point, the velocity along Z direction is approximately zero, and the robot starts moving along the X direction. This makes the robot perform the rotation movement of the book. In order to avoid the sliding of the book during this stage, the pushing force is kept constant. As the book is tilted, the tactile-based grasp controller updates the finger orientation in order to keep a suitable contact. Figure 8.13 shows a sequence in which the tactile controller establishes a wide contact area starting from a initially weak contact. Step 3 corresponds to the point where the hand adopts a new posture in order to grasp the book. Finally, during the fourth phase, the velocity reference is set to constant values, and the book is actually extracted from the bookshelf. The positive force that appears during this step is due to the book's weight. Figure 8.12b shows the trajectory followed by the fingertip during the execution of the whole task, represented within the XY plane of the robot's base coordinate system, which is aligned with the horizontal plane.

8.5 The UJI Librarian Robot

The book grasping ability is part of the more specialized UJI Librarian Robot. Nowadays, there is a great effort in robotics research to make robots capable of working in everyday scenarios, co-existing with humans. Currently, mobile service robots can be found in relatively unstructured environments such as museums, hospitals, etc. They typically make use of the available state-of-the-art technology to combine sensor-based navigation, localization and mapping, as well as obstacle avoidance capabilities and advanced user interfaces. However, service applications that include sensor-based manipulation in human environments are still very scarce. We believe that libraries offer a perfect start point to develop such applications, since they are human scenarios that are both ordinary and semi-structured. In addition, book grasping is one of the most important assistance actions that elderly and impaired people require [174, 65]. In this section, we present the UJI Librarian Robot, a system conceived not only for navigating in an ordinary library searching for a book, but also for autonomously extracting it from the bookshelf with a robot arm, and bringing it to the user.

Related previous work at Tsukuba University [184] included a tele-operated system aiming also at providing remote book browsing. In the authors' opinion, two important tasks needed improvement: optimization of the manipulation planning and book recognition; these topics are addressed in our research and satisfactorily solved. To the best of our knowledge, there is only one active project in the world that follows a goal similar to ours; namely, the Comprehensive Access to Printed Materials (CAPM) project [178]. Its main objective, presently under progress, will be to enable real-time browsing of printed materials through a web interface. An autonomous mobile system has been developed to retrieve items from bookshelves and carry them to scanning stations located in an off-site facility. In the current prototype, the books are stored in cases with a special design in order to facilitate grasping to the robot.

The UJI librarian robot is a complete robotic system designed to assist users in the library at Universitat Jaume I (UJI). The general functionality is illustrated in Figure 8.14. The user requests a book by using a modification of the standard web-based library interface, and the robot is then in charge of locating the book in the library, retrieve it and carry it to an assigned position. The system is only provided with the book code, which appears written on a label on each book spine. Using a vision system, it finds the particular bookcase and bookshelf where the book is, and integrates vision, force and tactile sensing to locate the target book and grasp it with a three-fingered Barrett Hand. In our approach, the environment is not modified for making things easier: books are not placed into special boxes for easier retrieval, we try to grasp the books as they can be found in current libraries, without making strong assumptions.

The application is composed of four different modules: a *user interface* module which also allows remote supervision, a *book localization* part, composed of intelligent book search algorithms and robot navigation, a *vision module* that performs the segmentation of labels in the image and processes them through Optical

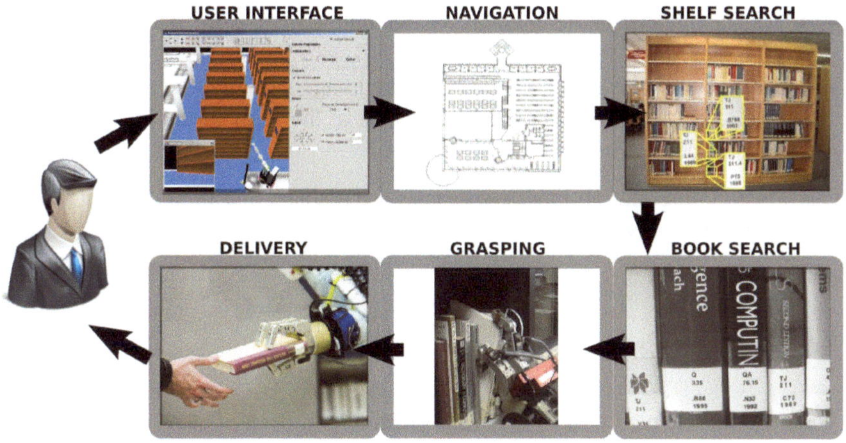

Fig. 8.14 The general approach of the UJI Librarian robot application

Character Recognition (OCR), and a *book retrieval* module that actually grasps the book from the bookshelf using the book grasping ability described in the previous section. Here, we focus on the vision algorithms. For a complete description of the rest of the application, please refer to [149, 150].

The vision module must be able to read book codes from single images, but book tracking is also necessary in order to keep a list of the labels that have been already processed by the OCR. We use a probabilistic tracking method in order to match labels through time, exploiting the spatio-temporal constraints of the problem. Using vision, book labels are segmented and located in the image. Next, they are processed by the OCR in order to read the identification codes. From the readings, the book localization module computes where to move the manipulator so that the book is found as soon as possible. The purpose of the tracking is to keep all labels located at any time, so that if the OCR module fails to recognize one of them reliably, it can be tried again later, from another point of view. In order to achieve the vision task, we propose a sequential and modular method composed of these steps: plane estimation, label segmentation, probabilistic matching, and optical character recognition.

The following constraints and assumptions about the problem are adopted:

- We assume that the camera is adequately placed in front of the shelf, so that the camera plane is approximately parallel to the book plane. This can be accomplished by estimating the book plane by means of vision, as it will be explained in section 8.5.1.
- All labels have white background and black foreground. In addition, they contain four lines of text, following the *US Library of Congress Classification* (LCC).
- Labels are all placed more or less at the same height in the book spine.
- Labels cannot overlap. This is not an assumption but a property of the problem that can be used to better detect regions.
- Labels are rectangular. The vertical size is approximately the same for all. The width depends on the width of the book.

- All the characters on the label are visible on the book spine, i.e. the book is not too thin.
- The environment is static. Changes on the image are only due to camera movements. Labels can only appear or disappear at the sides of the image, when the camera velocity is non zero.

It is worth noting that all these hypothesis are realistic and in agreement with our particular library environment.

8.5.1 Plane Estimation

When the mobile manipulator arrives at a position in front of the bookcase, the eye-in-hand camera must be aligned with the book plane, in order to have an optimal view of the books and make the image processing easier. As robot localization is not enough to ensure that this ideal position has been reached, we use vision for computing the book plane and automatically aligning the camera with it.

This step is currently done open-loop, by estimating the book plane once, at the initial camera position. It could also be performed in closed-loop by following a visual servoing approach. However, in this case we would need to select a set of stable visual features and to be able to track them in the image at each iteration. Precision in positioning would be better, but the execution would not be as reliable as open-loop. Even if making use of robust features and robust trackers, it would be very difficult to track the whole set of features during the alignment motion, even more knowing the reflective nature of most book surfaces. There is also the possibility to extract a whole set of new features at each iteration and computing the correspondences on the stereo image. However, this approach would be computationally expensive. In both cases, robot velocity should be small during the process, thus increasing the book delivery time. Open-loop precision was experimentally shown to be valid for this particular application. Thus, the minimalist solution was adopted, avoiding tracking and allowing fast alignment motion.

We use the well-known pin-hole camera model and epipolar geometry for linking the information contained in 2D images with the 3D structure of the projected world [52].

As the librarian robot has an eye-in-hand stereo camera, two different views are taken automatically. Both lenses point in the same direction, but are separated by a baseline of 90mm. We take the camera frame, C, at the middle point between both lenses, and another frame is attached to each lens, I and D, as shown in Figure 8.15.

Thus, taking C as the world center, the 3D coordinates of the reconstructed points will be defined with respect to this system. By the homogeneous transformations ${}^{I}\mathbf{M}_C$ and ${}^{D}\mathbf{M}_C$, both views are also defined in the common frame C. Thus, the projection matrices can be computed with the following expressions [52]:

$$\mathbf{P} = \mathbf{C}^{i}\mathbf{I}^{I}\mathbf{M}_C \tag{8.1}$$

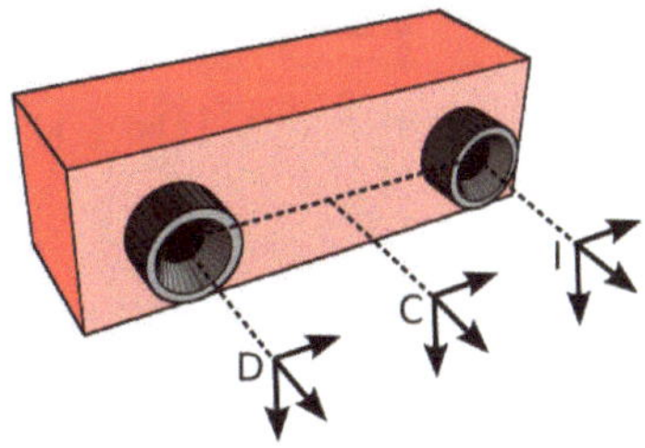

Fig. 8.15 Stereo camera and attached frames

$$\mathbf{P}' = \mathbf{C}^d \mathbf{I}^D \mathbf{M}_C \tag{8.2}$$

where $\mathbf{C}^i$ and $\mathbf{C}^d$ are the calibration matrices for each camera, and $\mathbf{I}$ is:

$$\mathbf{I} = \begin{pmatrix} 1\,0\,0\,0 \\ 0\,1\,0\,0 \\ 0\,0\,1\,0 \end{pmatrix}$$

Knowing $\mathbf{P} = (\mathbf{P}_r \quad \mathbf{p}_t)$ and $\mathbf{P}' = (\mathbf{P}'_r \quad \mathbf{p}'_t)$, the fundamental matrix, $\mathbf{F}$, can be easily computed by using the following well-known result from epipolar geometry [52]:

$$\mathbf{F} = \left[\mathbf{p}'_t - \mathbf{P}'_r \mathbf{P}_r^{-1} \mathbf{p}_t\right]_\times \mathbf{P}'_r \mathbf{P}_r^{-1} \tag{8.3}$$

Once $\mathbf{F}$ is known, correspondences can be found in a fast and robust way by looking for SIFT features in one of the images, and applying the sum of squared differences (SSD) method along the epipolar line in the other image. Having the correspondences, the 3D structure of the world can be computed. Figure 8.16a shows two images taken by the stereo cameras and a set of nine correspondences and three epipolar lines. Due to the particular arrangement of the cameras, both views point in parallel directions and the epipolar lines are horizontal. As books have normally different colors and text on their sides, the images are well-textured and it is easy to find many correspondences. In order to fit a plane, at least three 3D points must be reconstructed. The plane is computed by using standard least squares minimization. The greater the number of correspondences, the better the adjustment will be. Figure 8.16b shows the plane that is obtained from the correspondences. In the current implementation, outliers are not taken into account, making the plane estimation quite sensible to errors in the feature matching process. The use of robust estimators is one of the aspects that will be addressed in future work. The book plane could also be estimated by following a dense matching approach, and trying to fit a plane to a cloud of 3D points, but this method would take more computational time.

Having the plane equation in the camera frame, C, and knowing the relationship between the camera and the end-effector frame, ${}^E\mathbf{M}_C$, an open-loop end-effector displacement can be easily computed so that the camera is placed perpendicular to the plane, as it is desired for vision and grasping algorithms to work properly.

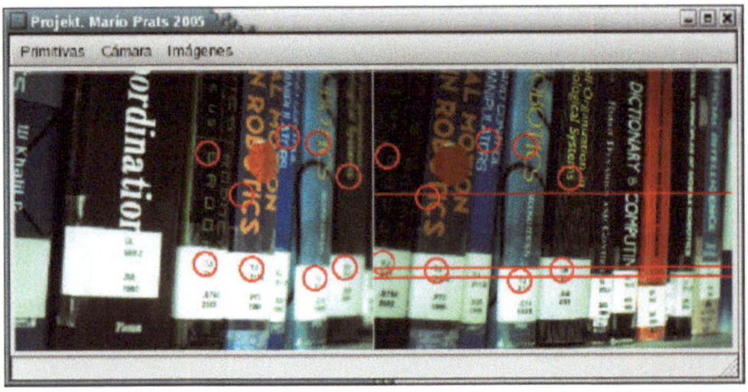

(a) A set of correspondences and epipolar lines in the stereo image

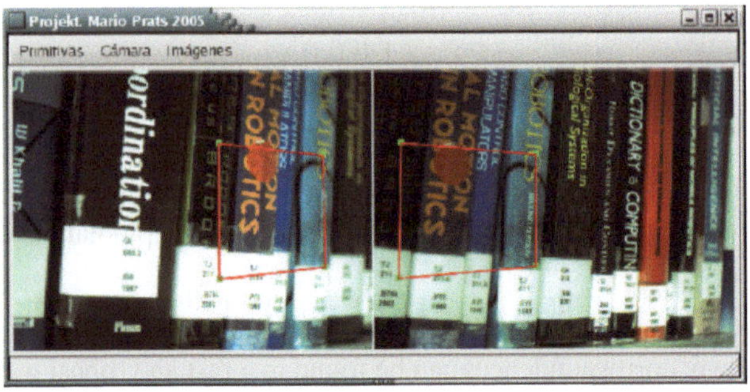

(b) The estimation of the book plane

Fig. 8.16 Estimation of the book plane from the image correspondences

8.5.2 Segmentation

Once a frontal view of the books is available, the label segmentation process starts. For this, we propose a local segmentation algorithm with automatic threshold detection. For a $N \times M$ greyscale image I, we first divide it into L vertical intervals of $C = \frac{M}{L}$ columns each one, as shown in Figure 8.17. We define the luminosity of the row i of interval k as:

$$l_k(i) = \frac{\sum_{j=k \cdot C}^{(k+1) \cdot C} I_{ij}}{C} \tag{8.4}$$

that is, the mean intensity of the ith row in interval k , being I_{ij} the value of the pixel with column j and row i (in the range $[0,255]$). Then, we find the row with more luminosity for each interval k:

$$L_k = \mathbf{max}\{l_k(i)\} \quad \forall i \in [0, N-1] \tag{8.5}$$

$$LR_K = \mathbf{argmax}_i\{l_k(i)\} \quad \forall i \in [0, N-1] \tag{8.6}$$

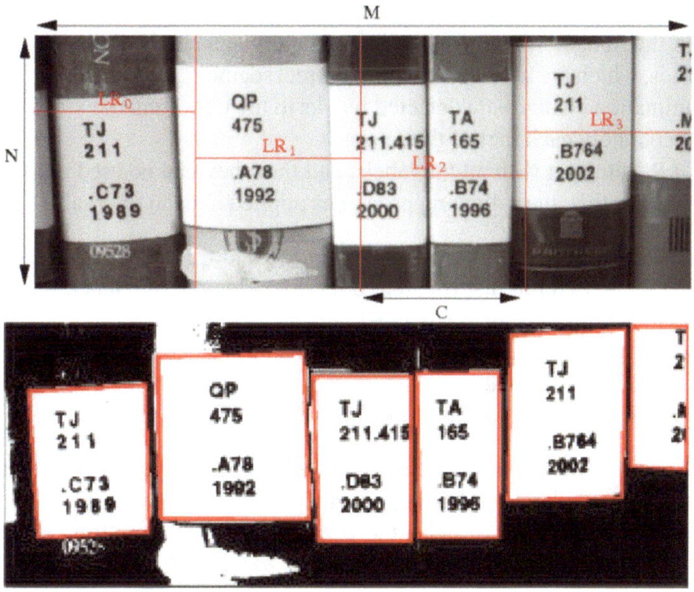

Fig. 8.17 Intervals defined on the image, and boundaries of the segmented labels

Horizontal lines in rows LR_k will normally intersect with book labels, with the exception of image intervals where all the books are white. In these cases, the row with more luminosity, LR_k, could not pass through the label. However, this is not a major problem because, since book and label are of the same color, we can treat it as a large label and refine the detection in later stages. Then, we can say that L_k corresponds to the mean intensity of labels.

The mean intensity of the row with more luminosity, L_k, can be used for segmentation purposes. We know that a book label is the lightest region we can find in the image, and its color value is around L_k. We can segment locally each interval k using a threshold value of $\frac{3}{4}L_k$, chosen experimentally. The result will be a binary image where labels are visible, along with some regions of the same color, as shown in Figure 8.17. Of course, the quality of the segmentation depends on the total number of intervals. For just one interval taking all the image, there is one L_0 that is the mean color of all the labels in the image. However, there are some labels that are darker than others, because they are older or due to shadows. In this case, L_0 would not be a good approximation of label intensity for all of them, and segmentation would not be precise (note that this would be the equivalent to a global segmentation). This problem is solved by taking more intervals into account and by segmenting the image locally. Of course, the labels of two books are more probable to have a similar intensity than those of six books. Then, higher values for L give a better segmentation, at the expense of lower execution speed.

Also note that the automatic selection of the threshold makes the segmentation algorithm robust to illumination conditions and illumination changes. Concretely, robustness in global illumination is due to the assumption that labels are the lightest regions in the image. The algorithm automatically detects the lightest rows and its mean values, which are used to estimate a correct segmentation threshold. Moreover, local illumination changes are detected thanks to the use of intervals which allow to segment separately each part of the image.

Rows LR_k intersect with all the labels, and thus they can be used to get the label boundaries. For this, the following process is applied to each interval $k \in [0, L-1]$:

1. Consider the row with more luminosity, LR_k, and find in the segmented image the next white pixel in that row that does not belong to any label, i.e., it has not been assigned previously to any label.
2. That pixel belongs to a new label, and we can use a standard contour extraction algorithm to get the boundary.
3. Repeat 1. and 2. until the end of the interval.

Finally, a curvature analysis on the boundaries is done in order to detect angles and lines. Detected boundaries that do not contain two pairs of parallel lines are rejected. Figure 8.17 shows the final result of the segmentation algorithm for a real image.

8.5.3 Probabilistic Matching, Motion Estimation and Selection

In order to extend our system to work over sequences of images, we can apply the segmentation process to each of them. But if we also want to follow the movement of a certain book, then we need a tracking method. In this application, tracking is especially interesting because we want to keep a list of the books that have already been successfully processed by the OCR module. For example, if after applying OCR to all the books of an image, the camera moves in the lateral direction so that a new book appears, then only the new label should be processed by the OCR module, and not the already processed ones.

We propose a model-based probabilistic approach [103, 48] for matching labels in two consecutive images. Rather than using the location of the labels in the first image to predict its location in the second image and avoid a whole segmentation, we take advantage of the good performance of the segmentation algorithm and execute a whole segmentation on each iteration. With this, we aim at improving the localization of the labels following a probabilistic framework. For example, segmentation is not always perfect, and it is possible to start with a low quality image of the label. Then, we apply segmentation to all the images while the camera is moving. If the label is segmented correctly in one image, our probabilistic algorithm will remember its features for helping future segmentations.

The matching algorithm tries to link books detected in the current image with those detected in the last one. For this, a suitable representation of a signature is needed. We consider that a label is fully described by its four vertex, represented as points **a**, **b**, **c**, **d** in the two dimensional image space. Points are ordered so that **a** corresponds to the top-left corner, **b** to the top-right, **c** to the bottom-right and **d** to the bottom-left. We can describe a label as a set of four vertex $\mathscr{S}_i = \{\mathbf{a}_i, \mathbf{b}_i, \mathbf{c}_i, \mathbf{d}_i\}$.

Following with the notation, signatures segmented at a given time t, are represented by the set $\mathscr{S}^t = \{\mathscr{S}_1^t, \mathscr{S}_2^t, \ldots, \mathscr{S}_n^t\}$, where n is the total number of books in the image. From this, the algorithm tries to match elements of $\mathscr{S}^t$ with elements in $\mathscr{S}^{t-1}$.

The matching is done following a probabilistic approach. Given two labels, $\mathscr{S}_i$ and $\mathscr{S}_j$, we define a metric for measuring the distance between them as:

$$M(\mathscr{S}_i, \mathscr{S}_j) = \frac{d_{\mathbf{a}_i\mathbf{a}_j} + d_{\mathbf{b}_i\mathbf{b}_j} + d_{\mathbf{c}_i\mathbf{c}_j} + d_{\mathbf{d}_i\mathbf{d}_j}}{4} \tag{8.7}$$

where d_{xy} is the point-to-point distance between points **x** and **y**. We also define the random variable X="*distance between two corresponding labels in two consecutive images*", and then:

$$X \hookrightarrow N(v_I, \sigma^2) \tag{8.8}$$

i.e, X is distributed by a Normal with mean v_I and variance σ^2. The mean v_I is an estimation of the image velocity and the variance is an estimation of the error in the velocity. The distance between the same signature in two consecutive images depends on the image velocity v_I, that can be estimated either by a calibrated camera with known motion, or by computing the image flow like we do. The variance in the distribution model can be chosen experimentally.

The system works by computing the distance of all possible matches of signatures between $\mathscr{S}^t$ and $\mathscr{S}^{t-1}$. Then, the probability model of expression 8.8 is applied in order to find the most probable pairs. Three cases can occur:

1. A label in $\mathscr{S}^t$ does not match with any label in $\mathscr{S}^{t-1}$. This can happen in two situations. The first one is at the beginning of the process, when $\mathscr{S}^{t-1}$ has no elements. The second one is when a new signature appears in the image, due to camera motion. In both cases, we assign an internal identification number to the book, that will be used for tracking.
2. A label in $\mathscr{S}^{t-1}$ does not match with any label in $\mathscr{S}^t$. This means that the signature has disappeared. As we have an estimation of the image flow, we can apply some restrictions about the place where signatures can disappear. For example, for a static image ($v_I = 0$), no signatures can disappear. However, if we displace the camera to the right, then signatures can only disappear through the left side of the image. Our system takes into account these constraints. If a signature disappears where it is not possible to, it is created again in $\mathscr{S}^t$ and displaced according to the prediction of the velocity. In this way, we can deal with occlusions or with errors in the segmentation process. The motion of lost labels is predicted knowing the motion of those which are matched.
3. A label $\mathscr{S}_i^t \in \mathscr{S}^t$ is matched with a label $\mathscr{S}_j^{t-1} \in \mathscr{S}^{t-1}$. That means that $\mathscr{S}_i^t$ in the current image is the same that $\mathscr{S}_j^{t-1}$ in the last image, so we have successfully matched the signature. But once more, segmentation problems can arise and it is possible to have a bad extraction in $\mathscr{S}_i^t$, while $\mathscr{S}_j^{t-1}$ is good. To cope with this

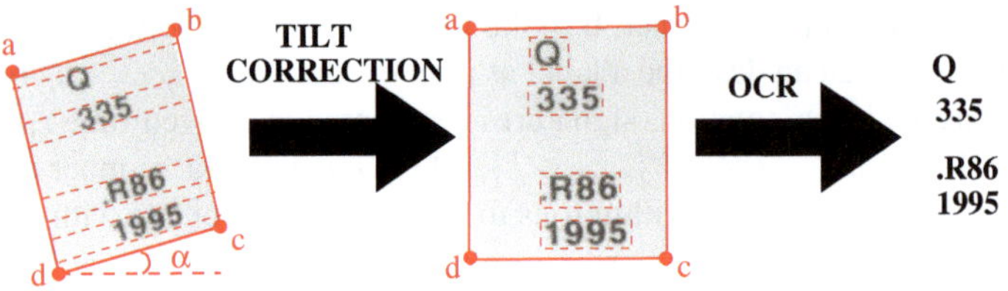

Fig. 8.18 The OCR process

problem, we define two additional random variables: H="*height of a signature*" and W="*width of a signature*". These variables are also distributed according to a Normal distribution, with a mean and a variance that can be chosen experimentally. When two labels match, the probability models are computed on $\mathscr{S}_i^t$ and $\mathscr{S}_j^{t-1}$. If $\mathscr{S}_i^t$ has a very low probability of being a good signature, and $\mathscr{S}_j^{t-1}$ has high probability, then we overwrite $\mathscr{S}_i^t$ with the good one and displace it in the image by using the image flow estimation. With this we ensure that if a label is well detected at a given time, future problems with it will be solved using the good prediction.

A good match is used to update the image flow in order to apply it in future predictions. The image flow is a 2D vector, $\mathbf{v}_I$, computed as follows:

$$\mathbf{v}_I = \frac{\mathbf{t_a} + \mathbf{t_b} + \mathbf{t_c} + \mathbf{t_d}}{4} \tag{8.9}$$

with:

$$\mathbf{t_a} = \mathbf{a^t} - \mathbf{a^{t-1}} \tag{8.10}$$
$$\mathbf{t_b} = \mathbf{b^t} - \mathbf{b^{t-1}} \tag{8.11}$$
$$\mathbf{t_c} = \mathbf{c^t} - \mathbf{c^{t-1}} \tag{8.12}$$
$$\mathbf{t_d} = \mathbf{d^t} - \mathbf{d^{t-1}} \tag{8.13}$$

The velocity used in the probability distribution is defined as the module of the image flow, $v_I = |\mathbf{v}_I|$.

Finally, the probability model of a label is used to eliminate those labels with a low probability of being amenable for grasping. With this, we can discriminate books that are too wide for being grasped by the robot, for example.

8.5.4 *Optical Character Recognition*

The recognition module takes as input a set of labels and it first tries to locate text lines inside them. For each label, we take parallel lines, perpendicular to the segment that joins points **a** and **d** (see Figure 8.18), and compute the luminosity for each line in a way similar to what is done with the full images. The maxima of this function represent the background while the minima represent text lines. In this way, the four lines of text are segmented. After that, the angle of the label is computed as:

$$\alpha = \arctan \frac{\mathbf{a}_y - \mathbf{b}_y}{\mathbf{b}_x - \mathbf{a}_x} \tag{8.14}$$

A rotation of $-\alpha$ is applied to the label image before it is passed to the OCR, in order to transform tilted labels into vertical ones. The utility of locating the separation between text lines is that each line can be passed separately to the OCR, and then we can apply some restrictions about the characters that can appear in each line. For example, we know that the first line of text should only contain letters, while the fourth line is a year, i.e., only numbers. If we process the lines separately, we can use these constraints on the OCR in order to reduce errors.

As OCR core, we use the GOCR software [167] that is being developed under the GNU Public License. This software is able to extract the text included in images. It works with all kind of fonts, and can be trained for better performance in particular applications. Once the desired book has been recognized, the grasping ability described in section 8.4.2 is executed.

8.6 Towards a Multipurpose Service Robot

Whereas the previous sections have described specific physical interaction capabilities, in this section we present the first steps towards a multipurpose assistive robot using the methods presented in this book. A service robot must be versatile, able to perform a wide range of tasks, and not only those for which it has been specifically programmed. Although most of our daily physical interaction can be classified into a few task categories, there is a huge number of different objects over which these tasks can be applied, thus making each manipulation experience unique. For example, the task of pulling open a door can be very different depending on the door mechanism, the handle shape, friction, etc. In addition, the vision-based estimation of the door pose also depends on the object physical properties, like color, texture or edges. A multipurpose robot must be able to deal with all these variants by implementing general vision-based methods and general manipulation approaches.

8.6.1 Vision-Based Localization of Objects

The use of object models allows to obtain a precise vision-based pose estimation in a general manner, in contrast to appearance-based methods where coarse approximations are normally obtained [127]. The physical interaction planner and sensor-based estimation and control methods presented in this book make use of object models, in order to offer the versatility and accuracy required for multipurpose physical

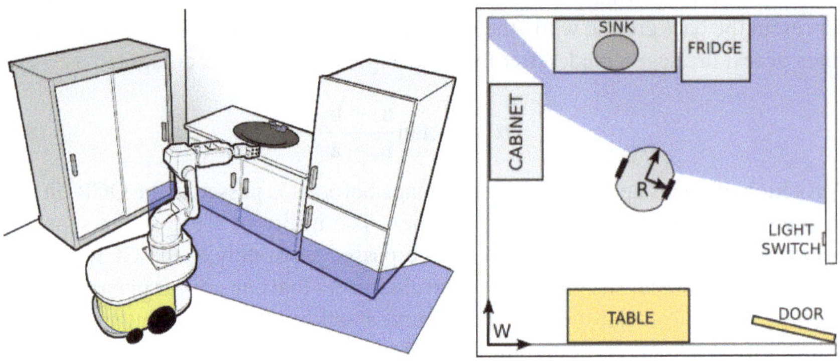

Fig. 8.19 A mobile manipulator in a kitchen environment. Map-based robot localization can be used for getting an initial estimation of the objects pose.

interaction. However, these models include very basic information, such as the box-based geometry or important edges, which can be acquired with current state-of-the-art vision-based approaches [75].

An important drawback of model-based methods is that a suitable initialization is needed, which is normally provided by the user [44], or through the previous application of an appearance-based method [47]. The first case is not suitable for autonomous manipulation, where the main goal is to avoid human intervention, whereas the second case requires an additional database containing the objects appearance from multiple points of view. However, it is also possible to obtain a first approximation of the objects pose from an environment map where the robot is already localized.

In general, it can be assumed that a service robot is always localized in the environment where it operates, as illustrated in Figure 8.19. There are many different robot localization approaches, and their accuracy depends on the type of sensors used for measuring the environment features: sonar [43], laser [21], robot vision [168], ceiling cameras [18], etc. Using one or a combination of these methods, the robot can be localized quite precisely inside a map of the environment, which can be acquired previously, or simultaneously to the localization process [46].

A considerable number of the objects we continuously manipulate in our daily life are fixed in the environment, or their position is constrained inside a limited space, such as doors, drawers, light switches, etc. Their pose in the environment can be known by the service robot, either because it is manually indicated by the user inside the environment map, or because the robot learns it during an exploration step [186], or through human demonstration [58]. Therefore, it can be assumed that an assistive robot can navigate close to these objects and be localized with respect to them with a precision of a few centimeters [142]. In the case of non-fixed objects, additional vision-based recognition methods would be needed in order to obtain an initial pose estimation, but previous knowledge about its location is normally available (e.g. on the table, inside the cabinet, etc.), thus reducing the search space.

Fig. 8.20 Application of the vision-based pose estimation and physical interaction planner to real cases

After an initial estimation of the target object pose is available, the VVS-based pose estimation approach described in section 6.2.2 can be applied in order to improve robot localization and obtain sub-centimeter accuracy, suitable for enabling manipulation. Figure 8.20 shows real images of common everyday objects where an initial error has been manually introduced in the real pose, in order to simulate

Fig. 8.21 Visual tracking of an oven along a video sequence

localization inaccuracies. The first column shows the initial pose estimation relying only on localization information. The second column shows the first iteration of the VVS minimization approach, and how a search is performed along the perpendicular directions of the estimated edges, looking for the real ones. The third column displays the estimated pose after the minimization has converged.

If the object is part of another object with bigger dimensions, the active pose approach proposed in this book can be applied in the parent object in order to obtain accurate information of the child object pose. The same pose estimation methods can be applied for object tracking, as shown Figure 8.21, where the oven pose is tracked during a video sequence. Therefore, it would be possible to perform a physical interaction task at the same time that the robot head is moving in order to accomplish a secondary task, such as improve the visibility, avoid occlusions, reach a comfortable posture, etc.

8.6.2 General Physical Interaction

The fourth column of Figure 8.20 shows the result of the physical interaction planner, superposed on the real images according to the pose estimation. A suitable hand preshape is selected for each particular case, and the physical interaction frames are set accordingly. The planning of the physical interaction task can be done from a high level description in terms of *object tasks* and *object actions* described in section 4.4. In the last two situations of Figure 8.20, different object actions are illustrated: opening the door and turning a knob for an oven, and switching on and selecting the temperature in the case of the washing machine.

Regarding control, the vision-force-tactile method described in section 7 can be adopted as a general approach. A particular sensor loop can be activated or deactivated depending on whether it provides valid information or not. For example, the vision loop should be deactivated during occlusion, or under poor lighting conditions. The particular control references have an important dependence on the task, and, thus, finding a general value suitable for all the cases is a difficult problem. An experience-based learning framework seems the most suitable approach for dealing with this problem, and it represents one of the immediate future extensions of the work presented in this book. Another pending issue is the detection of task-related events and the automatic planning of task transitions or recovery actions in case of failure. This is also highly task-dependent and, thus, it should be addressed from a robot learning point of view.

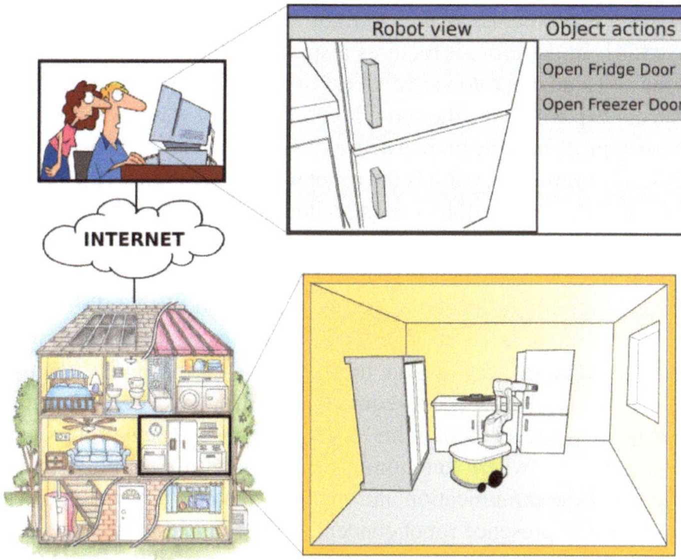

Fig. 8.22 An illustration of the remote physical interaction concept

8.6.3 Remote Physical Interaction

Tele-presence is one of the most important challenges in the tele-robotics community. The possibility to be present in a remote environment offers a wide range of very interesting applications, such as remote meetings, home supervision, surveillance of buildings, space exploration, home assistance, etc. Tele-presence systems aim at extending our sensor and motor capabilities over the network, potentially to any remote environment.

Although the concept of tele-presence involves the capabilities of perceiving and manipulating a remote environment, manipulation capabilities in current tele-presence systems are still very limited. Normally, a tele-presence system is composed of a mobile robot with a camera, speakers and microphones. Thus, it is possible to explore, see, talk and listen on remote spaces where we are not physically present. However, the capability of not only exploring a remote environment, but also interacting with it, introduces new interesting possibilities which could generate a great impact on the society. Just to mention a few examples, it would be possible to remotely close the windows at home in the case it starts raining, switch off a light that we left on, checking if there are enough aliments for the dinner before arriving home, etc. Tele-presence does not necessarily involve that the operator is far from the robot. In fact, operating a robot that is in our own environment would also be very useful, specially for people with mobility problems [136].

In the literature, attention has been focused on directly controlled tele-manipulation, where the human is inside the manipulation control loop, using a haptic device for controlling each robot arm movement [116]. However, this approach introduces several problems, as the operator fatigue, or how to deal with network time delays [54]. In addition, it requires a special equipment on the user side. The current trend is to advance towards supervisory control [169], or, intermediately, shared control [29], where the the robot is endowed with some autonomous capabilities. In these approaches, the human acts as a supervisor of the robot actions, giving some high-level commands, but it is the robot which decomposes the high-level task description into low-level actions which are locally controlled. The problems of direct control are thus eliminated. Firstly, the user does not require special equipment such as haptic devices. Instead, a web-based interface or special application can be used in order to send high-level commands to the robot. Secondly, as the robot is controlled locally, there is no need to transfer time-critical sensor signals through the network link, thus overcoming the time delay problem. Finally, the user fatigue is considerably reduced, as it is not required to control the low-level robot motion.

From the methods developed in this work, we envisage new promising advances towards tele-presence, where autonomous manipulation capabilities are added to the existing vision, communication and mobility factors. Figure 8.22 shows an illustration of our tele-presence robot concept. A user in a remote place receives into a computer or mobile device a map of the environment where the robot is operating, and real-time video captured with the robot camera. The user can remotely control the robot in order to explore the environment, as in classical tele-presence systems. As the robot is localized in the environment, it displays a list of the objects inside its camera view, and the actions that can be performed on them. However, the user is also allowed to command *object actions* that the robot autonomously transforms into physical interaction tasks, and performs by means of the sensor-based controllers, with the human supervision, but without its intervention. In this way, the robot can be remotely controlled at a high-level in order to manipulate doors, drawers, windows, home appliances, etc.

8.7 Discussion and Conclusions

This chapter has presented the first steps towards the application of the different contributions of this book in real service tasks. More concretely, the UJI Service Robot has been introduced, which is a first prototype of a mobile manipulator designed to assist in everyday tasks.

We have first defined a hybrid reactive-deliberative software architecture, where the different sensor-based controllers have been integrated as control actions, together with the physical interaction planner, which constitutes a perceptual action. The execution of the planning perceptual action automatically updates the perceptual state of the robot, which is used as input to the control actions. Perceptual and control actions are grouped into *abilities*, which define high-level execution plans. Two abilities have been created: the door opening and the book grasping abilities, being the latter part of the more specialized application of the UJI Librarian Robot.

The architecture allows for the dynamic creation of abilities by high-level task planning algorithms, implemented as behavioral actions. Although this property has still not been exploited, it represents a basic capability for autonomous robots. For example, a high-level task planner could dynamically create a specialized ability for a given physical interaction task, including the physical interaction planner, suitable sensor-based control actions, and the corresponding sensor-based conditions linking them.

Finally, the vision-based pose estimation methods have been applied to real situations that a service robot may encounter during its daily operation, and a 3D model of an *ideal hand* has been rendered on the real images, following an augmented reality approach, and based on the specification generated by the physical interaction planner. The number of different real situations where these methods have been applied foresees a promising future in our way to add versatile manipulation into assistive robots.

Unfortunately, building a general sensor-based controller is still a challenging problem far from being solved. Although the vision-tactile-force controller presented in section 7 provides a reliable and versatile solution for the execution of physical interaction tasks, it needs vision, tactile and force references which depend on the particular task. Additional methods for the automatic setting of control references are needed, and robot experience-based learning approaches may play an important role here.

Chapter 9
Conclusions

This chapter concludes the book, by remarking the main achievements of our research, and mentioning a number of future lines that have not been addressed throughout the document.

In this book, the new concept of physical interaction has been introduced as a unified representation of grasp and task-related aspects, which have been normally addressed from different point of views. The main purpose of the integrated treatment of the grasp and the task is to increase the versatility, autonomy and dependability of current manipulation approaches. In fact, the consideration of task-related aspects in the selection of a grasp extends the robot capabilities beyond pick and place actions. Similarly, the control of grasp-related aspects during task execution increases the overall execution reliability, as deficient contact configurations can be detected and corrected.

With the purpose of gaining in versatility, autonomy and dependability, new methods for the specification, planning and control of physical interaction tasks have been proposed. The specification is done by means of the *physical interaction framework*, where both the grasp and the task are considered in an integrated manner. Planning is performed through the *physical interaction planner*, which selects the most suitable task-oriented hand posture for a specific high level task, given in terms of *object actions*. Finally, reliable execution is performed through the combination of vision, force and tactile feedback, according to the case. Concretely, new force, vision-force and vision-force-tactile control approaches have been proposed, which outperform the existing methods. All the developments have been implemented and validated in different robotic systems and considering several physical interaction tasks.

The integrated treatment of the grasp and the task, and its execution in terms of multiple sensor combination, introduces important advantages with respect to the state of the art in robot-environment interaction:

Versatility. The physical interaction framework supports a wide range of actions, not only involving prehensile grasping, but also non-prehensile manipulation and more complex tasks like using tools or two hands. The planner supports a subset

M. Prats et al.: Robot Physical Interaction through the Combination of Vision, STAR 84, pp. 159–165.
springerlink.com

Table 9.1 The different tasks and sensor combinations that have been applied with the robotic systems

Robot	**Hand**	**Task**	**Sensor**
Armar-III Humanoid	Five-fingered hand	Open fridge Open dishwasher Open drawer Open cabinet	Force
UJI Service Robot	Parallel jaw gripper	Grasp book	
	Barrett Hand	Turn handle	
		Grasp book	Force-tactile
		Open sliding door	Vision-force-tactile
ISRC Mobile Manip.	Parallel jaw gripper	Open closet	Vision-force
		Open fridge	

of these actions, and more concretely, those related with interaction with furniture and home appliances in household environments, such as turning handles, opening doors, pushing buttons, etc. Different robotic hands are also supported thanks to the implementation of *hand adaptors*. All the control methods rely on cartesian control of the robot end-effector, thus making its application into different manipulators possible. In fact, three different robots with three different hands have been used for validation of the proposed methods. Up to ten different physical interaction tasks have been executed, some of them replicated in different robots, as summarized in Table 9.1. Regarding the vision part, many different objects can be successfully tracked in real conditions, as it has been shown in section 8.6.1. In addition, articulated objects are also supported, and almost any kind of visual feature can be used and combined with the VVS approach (points, lines, ellipses, etc.), as long as the corresponding interaction matrix can be computed.
Dependability. Robustness to modeling errors and geometric uncertainties is achieved thanks to the combination of multiple sensors. Force sensors provide a confident signal reporting the interaction forces, whereas tactile sensors can detect contacts at the finger level very accurately. As it has been shown in chapters 6, 7 and 8, vision can report the pose of the objects in the environment, and, although its information does not present the reliability of the other sensors, robust estimators that are able to deal with outliers in the image processing have been implemented [25]. Even if the vision estimation is wrong or inaccurate, the integration with tactile and force controllers ensure a robust and safe manipulation, as reported in chapter 7. In the case where only force feedback is available, it is still possible to perform different constrained tasks without knowing the particular constrained trajectories, as shown in section 5.4.
Autonomy. The physical interaction planner of chapter 4 provides autonomous specification of tasks concerning interaction with articulated furniture and home appliances. The increased reliability of the execution methods also provides further autonomy at the control level, in the sense that human intervention in the

control loop is not required. However, reaching the levels of autonomy that potential users would require [136] is still a challenging problem. Additional methods for the automatic selection of suitable control references should be developed.

Viability. The sensor-based nature of the control approaches allow to perform physical interaction tasks in the real world, even with a very limited knowledge of the environment model, and under important uncertainties. The use of grasp redundancy, in addition to joint redundancy, allows to expand the workspace of the robot in applications which would be not possible to perform otherwise (section 5.4.3.5). Finally, the vision-based methods allow to use the natural object features for pose estimation, instead of special markers, as shown in chapters 7 and 8. In addition, active pose estimation can be performed in situations where the robot is close to objects that do not fit in the camera view, which is a common case in real scenarios, as it has been described in section 6.2.2. The models required for planning and visual tracking are composed of simple box-shaped primitives which can be easily built with state-of-the-art techniques [75].

Solid background theory. All the developments presented in this book are based on well-established concepts which have been widely accepted in the robotics literature, such as the Task Frame Formalism, the knowledge-based approach to grasping, stiffness force control, or the vision part, which relies on Virtual Visual Servoing. This method is formulated as a 2D visual servoing problem, for which convergence conditions, stability, robustness, etc. have been widely studied in the visual servoing community [76].

Efficiency Computational times are very important in closed-loop control approaches. Whereas force and tactile sensors normally provide data at a very high rate, vision processing usually requires additional computational time. However, the vision methods used in this book have shown to be very efficient in computational terms [26], allowing for processing at video rate. This has the advantage that the same methods used for pose estimation can be also used for pose tracking and for control purposes, even without using prediction models like Kalman filters.

9.1 Contributions

The main contributions of this book are the following:

- The concept of physical interaction task, as a global approach involving two sequential aspects of our interaction with the environment: the reaching action which establishes a set of contacts, and the task motion performed through these contacts.
- A framework for the specification of physical interaction tasks, based on well-established techniques adopted from the task control and grasping communities. The framework not only supports the classical grasping for pick and place tasks, but also other types of interaction. Complex tasks, such as using tools or

two-hand manipulation, can be described as a set of individual physical interaction tasks and, thus, specified through different instantiations of this framework.

- The new concept of *task-oriented hand preshape*, defined as an extension to the classical notion of hand posture, by including task-related aspects such as the hand frame.
- The concepts of *ideal hand* and *hand adaptors*, introduced as a way to define *ideal task-oriented hand preshapes* that can be instantiated on different real robotic hands.
- A planning algorithm that allows to automatically generate physical interaction tasks from a high-level description in terms of *object actions* and *object tasks*. The planner sets all the required elements that compose the specification of the grasp and the task on the basis of the physical interaction framework.
- A method for the force-based tracking of the physical interaction frames, allowing the robot to naturally adapt to the particular constrained trajectory without planning the specific motion trajectory.
- A force-based controller that integrates joint and grasp redundancy for increasing the robot workspace while performing constrained motion.
- An active vision approach for pose estimation of large objects which do not fit completely in the camera view. The method is based on the Virtual Visual Servoing approach, and requires several views of different object parts in order to acquire a rich set of visual features and build a full-rank interaction matrix. This method allows to recover the object pose in situations where the other approaches cannot be applied, as for example when the robot camera is so close to the object that only a small part of it is visible, which is a common case in manipulation environments.
- A new vision-force control scheme, called external hybrid vision-force control, that provides several advantages with respect to the existing methods.
- A new approach for integrating position, vision, tactile and force information into a hybrid controller, where the most confident available sensor information is selected for each cartesian degree of freedom. Force information, however, is used on all the directions in order to ensure a safe physical interaction.
- A hybrid reactive-deliberative software architecture where the different sensor-based controllers have been integrated, and which allows for the dynamic creation of abilities by high-level task planning algorithms.

These theoretical developments have lead to several practical contributions:

- A force-based method for grasping a book with a parallel jaw gripper, while standing among other books.
- Robust force-based interaction of the Armar-III Humanoid Robot (Karlsruhe University, Germany) with kitchen furniture, such as opening the fridge, dishwasher, cabinets and drawers, without relying on its geometric models.
- Implementation of an endpoint closed-loop external vision-force approach into the ISRC Mobile Manipulator (Sungkyunkwan University, Suwon, South Korea), in terms of the physical interaction framework, with the main advantage that the

camera can be moved freely without affecting the task execution, providing an approach amenable to be integrated into current humanoid robots without the need for external camera calibration.
- A sliding door opening task in a real environment by means of the combination of the control signals provided by a position controller, which has an initial coarse estimation of the object pose, a vision controller based on the articular object pose computed by the Virtual Visual Servoing method, a tactile controller, which looks for not losing the contact during manipulation, and a stiffness force controller, in charge of pushing along the task direction at the same time that the force is regulated on the rest of directions.
- Implementation of force-based door opening and force-tactile-based book grasping skills in the UJI Service Robot prototype.
- Vision-based methods for book recognition with the UJI Librarian Robot prototype.
- Application of the vision-based pose estimation methods and the physical interaction planner to real situations that an assistive robot may encounter during its daily operation, including objects like a washing machine, an oven, a light switch, a coffee machine, an elevator, etc.

9.2 Future Lines

The work that has been presented throughout this document leads to a large number of interesting issues that should be addressed in the future. Among them, we specially remark the following:

9.2.1 Purposive Use of Objects

The use of tools was one of the most important aspects in the human development. Although some guidelines about how to model the purposive use of objects with the proposed framework have been introduced in Chapter 3, no practical experiments have been still carried out showing its suitability. Advanced manipulation capabilities should include the intentional use of objects for achieving certain tasks. This includes the use of tools, like kitchen utensils, a screwdriver, a hammer, etc., but also other objects which are not considered as tools, but are needed for the success of some tasks, such as the insertion of keys, electronic cards, coins, and so on. In these objects, the grasp frame should be set so that the part required for the task remains available after the grasp has been performed. Therefore, additional planning methods should be developed for considering these issues.

9.2.2 *Platform Mobility*

We plan an evolution in the current version of the UJI Service Robot that includes new manipulation capabilities embodied in a renewed hardware, with the ultimate purpose to provide a multipurpose autonomous assistive manipulator. For this, it will be necessary to consider the mobility issue, which has not been addressed in this book. The framework and control strategies proposed in this document are based on the cartesian control of the end-effector, no matter whether this motion is generated only by the arm, or also by the rest of the robotic system. In order to integrate platform mobility, the additional DOF's should be integrated into the current cartesian controller. This would allow to increase the workspace of the robot and perform those tasks that require larger trajectories.

9.2.3 *Object Recognition*

Object recognition is the process by which a robot can identify the objects in its surroundings. It normally involves the assignment of a label that identifies the object uniquely, or as a member of a particular class. This label is typically used in order to access to a knowledge repository and obtain useful information, such as geometric models, functionality, physical properties, etc. Most of the objects that have been considered throughout this document were fixed in the environment, and, thus, easy to recognize from its position in the map. However, an assistive robot will also need to recognize smaller objects with non-fixed position, such as bottles, tools, etc. We plan to study different object recognition approaches, and specially those based on *Radio-frequency Identification* (RFID) [157], which is an emergent technology that could imply a breakthrough in recognition techniques.

9.2.4 *Automatic Setting of Control Parameters*

The execution of physical interaction tasks require setting suitable references in the corresponding control approaches. It is necessary, for example, to establish the maximum force to apply for turning a handle, a suitable opening angle for a door, the force to be applied when pushing a button, an appropriate pressure when grasping a delicate object, etc. It is not possible to set general values which are valid across different objects. Each object represents a particular case, and requires concrete references which cannot be known in advance. In our opinion, this is a problem which can be focused from a robot learning point of view. Exploration procedures must be defined so that the robot can autonomously acquire these object-specific values, and store them for future experiences.

9.2.5 Application to Autonomous Underwater Manipulation

Part of the developments described in this work have been applied to autonomous underwater intervention in the framework of the projects RAUVI (Reconfigurable Autonomous Unmanned Vehicle for Intervention, [160]), and FP7 TRIDENT (Marine Robots and Dexterous Manipulation for Enabling Autonomous Underwater Multipurpose Intervention Missions, [162]). These projects aim at developing underwater robots able to perceive the environment by means of acoustic and optic sensors, and equipped with a robotic arm in order to autonomously perform complex intervention tasks. The overall mission is divided in two steps: a *survey*, in which the robot obtains information about the underwater environment, and the *intervention*, where the robot performs the manipulation task based on vision, force and touch sensors. Figure 9.1 shows a sequence of the underwater robot recovering a flight data recorder.

Fig. 9.1 Experiments on autonomous black box recovery with the RAUVI prototype. From left to right: the arm is approaching the target while the vehicle performs station keeping. After retrieval of the target the vehicle returns to the surface.

References

1. Aiyama, Y., Inaba, M., Inoue, H.: Pivoting: A new method of graspless manipulation of object by robot fingers. In: IEEE/RSJ International Conference on Intelligent Robots and Systems, vol. 1, pp. 136–143 (1993)
2. Allen, P.K., Miller, A., Oh, P.Y., Leibowitz, B.: Integration of vision, force and tactile sensing for grasping. International Journal of Intelligent Machines 4(1), 129–149 (1999)
3. Alqasemi, R., McCaffrey, E., Edwards, K., Dubey, R.: Wheelchair-mounted robotic arms: analysis, evaluation and development. In: IEEE/ASME International Conference on Advanced Intelligent Mechatronics, pp. 1164–1169 (2005), doi:10.1109/AIM.2005.1511167
4. An, C., Hollerbach, J.: Kinematic stability issues in force control of manipulators. In: IEEE International Conference on Robotics and Automation, vol. 4, pp. 897–903 (1987)
5. Arkin, R.: Behavior-Based Robotics. MIT Press (1998)
6. Asfour, T., Regenstein, K., Azad, P., Schroder, J., Bierbaum, A., Vahrenkamp, N., Dillmann, R.: Armar-iii: An integrated humanoid platform for sensory-motor control. In: IEEE-RAS International Conference on Humanoid Robots, pp. 169–175 (2006)
7. Baeten, J., Bruyninckx, H., De Schutter, J.: Integrated vision/force robotic servoing in the task frame formalism. International Journal of Robotics Research 22(10-11), 941–954 (2003)
8. Bajcsy, R.: Integrating vision and touch for robotic applications. In: Reitman, W. (ed.) Trends and Applications of AI in Business. Ablex Publ. Co. (1984)
9. Bard, C., Laugier, C., Milési-Bellier, C., Troccaz, J., Triggs, B., Vercelli, G.: Achieving dextrous grasping by integrating planning and vision-based sensing. International Journal of Robotics Research 14(5), 445–464 (1995)
10. Bekey, G., Ambrose, R., Kumar, V., Lavery, D., Sanderson, A., Wilcox, B., Yuh, J., Zheng, Y.: Robotics: State of the Art and Future Challenges. Imperial College Press (2008)
11. Bicchi, A., Kumar, V.: Robotic grasping and contact: A review. In: IEEE International Conference on Robotics and Automation, San Francisco, CA, pp. 348–353 (2000)
12. Bicchi, A., Salisbury, J., Dario, P.: Augmentation of grasp robustness using intrinsic tactile sensing. In: IEEE International Conference on Robotics and Automation, vol. 1, pp. 302–307 (1989)
13. Bien, Z., Chung, M.J., Chang, P.H., Kwon, D.S., Kim, D.J., Han, J.S., Kim, J.H., Kim, D.H., Park, H.S., Kang, S.H., Lee, K., Lim, S.C.: Integration of a rehabilitation robotic system (kares ii) with human-friendly man-machine interaction units. Autonomous Robots 16(2), 165–191 (2004)

14. Borst, C., Fischer, M., Hirzinger, G.: A fast and robust grasp planner for arbitrary 3d objects. In: IEEE International Conference on Robotics and Automation, pp. 1890–1896 (1999)
15. Borst, C., Fischer, M., Hirzinger, G.: Grasping the dice by dicing the grasp. In: IEEE/RSJ International Conference on Intelligent Robots and Systems, Las Vegas, Nevada, USA (2003)
16. Borst, C., Fischer, M., Hirzinger, G.: Grasp Planning: How to Choose a Suitable Task Wrench Space. In: IEEE International Conference on Robotics and Automation, New Orleans, USA, pp. 319–325 (2004)
17. Borst, C., Ott, C., Wimbock, T., Brunner, B., Zacharias, F., Bauml, B., Hillenbrand, U., Haddadin, S., Albu-Schaffer, A., Hirzinger, G.: A humanoid upper body system for two-handed manipulation. In: IEEE International Conference on Robotics and Automation, pp. 2766–2767 (2007), doi:10.1109/ROBOT.2007.363886
18. Broxvall, M., Gritti, M., Saffiotti, A., Seo, B., Cho, Y.: PEIS ecology: Integrating robots into smart environments. In: IEEE International Conference on Robotics and Automation, Orlando, FL, pp. 212–218 (2006)
19. Bruyninckx, H., De Schutter, J.: Specification of force-controlled actions in the 'task frame formalism': A synthesis. IEEE Transactions on Robotics and Automation 12(5), 581–589 (1996)
20. Bruyninckx, H., De Schutter, J., Allotta, B.: Model-based constrained motion. a. modeling, specification and control. In: International Conference on Advanced Robotics, vol. 2, pp. 976–981 (1991)
21. Castellanos, J., Neira, J., Strauss, O., Tardos, J.: Detecting high level features for mobile robot localization. In: IEEE International Conference on Multisensor Fusion and Integration, Washington, DC, USA, pp. 611–618 (1996)
22. Chaumette, F.: Potential problems of stability and convergence in image-based and position-based visual servoing. In: The Confluence of Vision and Control. LNCIS, vol. 237, pp. 66–78. Springer (1998)
23. Chaumette, F., Hutchinson, S.: Visual servo control, part i: Basic approaches. IEEE Robotics & Automation Magazine 13(4), 82–90 (2006), `http://www.irisa.fr/lagadic/publi/publi/Chaumette07a-eng.html`
24. Chaumette, F., Hutchinson, S.: Visual servo control, part ii: Advanced approaches. IEEE Robotics & Automation Magazine 14(1), 109–118 (2007)
25. Comport, A., Pressigout, M., Marchand, E., Chaumette, F.: A visual servoing control law that is robust to image outliers. In: IEEE/RSJ International Conference on Intelligent Robots and Systems, vol. 1, pp. 492–497 (2003)
26. Comport, A.I., Kragic, D., Marchand, E., Chaumette, F.: Robust real-time visual tracking: Comparison, theoretical analysis and performance evaluation. In: IEEE International Conference on Robotics and Automation, Barcelona, Spain, pp. 2852–2857 (2005)
27. Comport, A.I., Marchand, E., Chaumette, F.: Object-based visual 3d tracking of articulated objects via kinematic sets. In: Computer Vision and Pattern Recognition Workshop, vol. 1, p. 2. IEEE Computer Society, Washington, DC (2004)
28. Comport, A.I., Marchand, E., Chaumette, F.: Robust model-based tracking for robot vision. In: IEEE/RSJ International Conference on Intelligent Robots and Systems, pp. 692–697 (2004)
29. Conway, L., Volz, R., Walker, M.: Tele-autonomous systems: Methods and architectures for intermingling autonomous and telerobotic technology. In: IEEE International Conference on Robotics and Automation, vol. 4, pp. 1121–1130 (1987)

30. Corke, P., Hutchinson, S.: A new partioned approach to image-based visual servo control. In: IEEE International Conference on Decision and Control, Sidney, pp. 2521–2526 (2000)
31. Cutkosky, M.: Grasping and fine manipulation for automated manufacturing. Ph.D. thesis, Carnegie-Mellon Univ., Dept. of Mechanical Engineering (1985)
32. Cutkosky, M., Hyde, J.: Manipulation control with dynamic tactile sensing. In: 6th International Symposium on Robotics Research, Hidden Valley, Pennsylvania (1993)
33. Cutkosky, M., Wright, P.: Modeling manufacturing grips and correlations with the design of robotic hands. In: IEEE International Conference on Robotics and Automation, San Francisco, CA, vol. 3, pp. 1533–1539 (1986)
34. Dario, D., De Rossi, D.: Tactile sensors and the gripping challenge. IEEE Spectrum 5(22), 46–52 (1985)
35. De Schutter, J., van Brussel, H.: Compliant robot motion - ii - a control approach based on external control loop. International Journal of Robotics Research 7(4), 18–33 (1988)
36. De Schutter, J., van Brussel, H.: Compliant robot motion: I. a formalism for specifying compliant motion tasks. International Journal of Robotics Research 7(4), 3–17 (1988), http://dx.doi.org/10.1177/027836498800700401
37. De Schutter, J., De Laet, T., Rutgeerts, J., Decré, W., Smits, R., Aertbeliën, E., Claes, K., Bruyninckx, H.: Constraint-based task specification and estimation for sensor-based robot systems in the presence of geometric uncertainty. International Journal of Robotics Research 26(5), 433–455 (2007)
38. Dementhon, D., Davis, L.: Model-based object pose in 25 lines of code. International Journal of Computer Vision 15(1/2), 123–141 (1995)
39. Diankov, R., Srinivasa, S., Ferguson, D., Kuffner, J.: Manipulation planning with caging grasps. In: IEEE-RAS International Conference on Humanoid Robots, pp. 285–292 (2008), doi:10.1109/ICHR.2008.4755966
40. Dionnet, F., Marchand, E.: Robust stereo tracking for space applications. In: IEEE/RSJ International Conference on Intelligent Robots and Systems, pp. 3373–3378 (2007), doi:10.1109/IROS.2007.4399028
41. Drake, S.: Using compliance in lieu of sensory feedback for automatic assembly. Ph.D. thesis. Dept. of Mechanical Engineering. MIT, USA (1977)
42. Driessen, B.J., Evers, H.G., van Woerden, J.A.: Manus–a wheelchair-mounted rehabilitation robot. Proc. Inst. Mech. Eng. [H] 215(3), 285–290 (2001)
43. Drumheller, M.: Mobile robot localization using sonar. IEEE Transactions on Pattern Analysis and Machine Intelligence 9, 325–332 (1987)
44. Drummond, T., Cipolla, R.: Real-time visual tracking of complex structures. IEEE Transactions on Pattern Analysis and Machine Intelligence 24(7), 932–946 (2002)
45. Dune, C., Marchand, E., Collewet, C., Leroux, C.: Active rough shape estimation of unknown objects. In: IEEE/RSJ International Conference on Intelligent Robots and Systems, Nice, France, pp. 3622–3627 (2008)
46. Durrant-Whyte, H., Bailey, T.: Simultaneous localization and mapping: part i. IEEE Robotics & Automation Magazine 13(2), 99–110 (2006)
47. Ekvall, S., Kragic, D., Hoffmann, F.: Object recognition and pose estimation using color cooccurrence histograms and geometric modeling. Image and Vision Computing 23(11), 943–955 (2005)
48. Elgammal, A., Duraiswami, R., Davis, L.: Probabilistic tracking in joint feature-spatial spaces. In: IEEE Conference on Computer Vision and Pattern Recognition, Madison, WI, USA, pp. 781–788 (2003)
49. Eltaib, M., Hewit, J.: Tactile sensing technology for minimal access surgery–a review. Mechatronics 13(10), 1163–1177 (2003)

50. Espiau, B., Chaumette, F., Rives, P.: A new approach to visual servoing in robotics. IEEE Transactions on Robotics and Automation 8(3), 313–326 (1992)
51. Evans, J.: Helpmate: an autonomous mobile robot courier for hospitals. In: IEEE/RSJ International Conference on Intelligent Robots and Systems, vol. 3, pp. 1695–1700 (1994), doi:10.1109/IROS.1994.407629
52. Faugeras, O., Luong, Q., Papadopoulo, T.: The geometry of multiple images. The MIT Press, Cambridge (2001)
53. Ferrari, C., Canny, J.: Planning optimal grasps. In: IEEE International Conference on Robotics and Automation, Nice, France, vol. 3, pp. 2290–2295 (1992)
54. Ferrell, W.: Remote manipulation with transmission delay. IEEE Transactions on Human Factors in Electronics 6, 24–32 (1965)
55. Finkemeyer, B., Kroger, T., Wahl, F.M.: Executing assembly tasks specified by manipulation primitive nets. Journal of Advanced Robotics 19, 591–611 (2005)
56. Gates, B.: A robot in every home. Sci. Am. 296(1), 58–65 (2007)
57. George, B.: Autonomous Robots. The MIT Press (2005)
58. Ghidary, S.S., Nakata, Y., Saito, H., Hattori, M., Takamori, T.: Multi-modal interaction of human and home robot in the context of room map generation. Autonomous Robots 13(2), 169–184 (2002)
59. Graf, B., Hans, M., Schraft, R.: Care-o-bot ii-development of a next generation robotic home assistant. Autonomous Robots 16(2), 193–205 (2004)
60. Grupen, R., Brock, O.: White paper: Integrating manual dexterity with mobility for human-scale service robotics – the case for concentrated research into science and technology supporting next-generation robotic assistants (2004)
61. Gunji, D., Mizoguchi, Y., Teshigawara, S., Ming, A., Namiki, A., Ishikawaand, M., Shimojo, M.: Grasping force control of multi-fingered robot hand based on slip detection using tactile sensor. In: IEEE International Conference on Robotics and Automation, Pasadena, CA, USA, pp. 2605–2610 (2008)
62. Gupta, K., del Pobil, A.P.: Practical Motion Planning in Robotics: Current Approaches and Future Directions. John Wiley & Sons, Inc., New York (1998)
63. Han, L., Trinkle, J.C., Li, Z.: Grasp analysis as linear matrix inequality problems. IEEE Transactions on Robotics and Automation 16, 1261–1268 (2000)
64. HAR: Online, `http://www.irt.i.u-tokyo.ac.jp/news/pressrelease081024_e.pdf`
65. Harmo, P., Taipalus, T., Knuuttila, J., Vallet, J., Halme, A.: Needs and solutions - home automation and service robots for the elderly and disabled. In: IEEE/RSJ International Conference on Intelligent Robots and Systems, pp. 3201–3206 (2005), doi:10.1109/IROS.2005.1545387
66. Haschke, R., Steil, J., Steuwer, I., Ritter, H.: Task-oriented quality measures for dextrous grasping. In: IEEE Conf. on Computational Intelligence in Robotics and Automation, Espoo, Finland, pp. 689–694 (2005)
67. Hasegawa, T., Suehiro, T., Takase, K.: A model-based manipulation system with skill-based execution. IEEE Transactions on Robotics and Automation 8(5), 535–544 (1992)
68. Hogan, N.: Impedance control of industrial robots. Robotics and Computer-Integrated Manufacturing 1(1), 97–113 (1984)
69. Horaud, R., Dornaika, F., Espiau, B.: Visually guided object grasping. IEEE Transactions on Robotics and Automation 14(4), 525–532 (1998)
70. Hosoda, K., Igarashi, K., Asada, M.: Hybrid visual servoing / force control in unknown environment. In: IEEE/RSJ International Conference on Intelligent Robots and Systems, Osaka, Japan, pp. 1097–1103 (1996)

71. Howe, R.: Tactile sensing and control of robotic manipulation. Journal of Advanced Robotics 8(3), 245–261 (1994)
72. Howe, R.D., Cutkosky, M.R.: Touch sensing for robotic manipulation and recognition. MIT Press, Cambridge (1992)
73. Hu, Y., Eagleson, R., Goodale, M.: Human visual servoing for reaching and grasping: The role of 3-d geometric features. In: IEEE International Conference on Robotics and Automation, Detroit, Michigan, USA, pp. 3209–3216 (1999)
74. Huebner, K., Kragic, D.: Selection of robot pre-grasps using box-based shape approximation. In: IEEE/RSJ International Conference on Intelligent Robots and Systems, pp. 1765–1770 (2008)
75. Huebner, K., Ruthotto, S., Kragic, D.: Minimum volume bounding box decomposition for shape approximation in robot grasping. In: IEEE International Conference on Robotics and Automation, pp. 1628–1633 (2008)
76. Hutchinson, S., Hager, G., Corke, P.: A tutorial on visual servo control. IEEE Transactions on Robotics and Automation 12(5), 651–670 (1996)
77. I-Sobot, `http://www.isobotrobot.com/`
78. IFR, S.D.: World robotics 2008 (2008), `http://www.worldrobotics.org/`
79. IFR, S.D.: World robotics 2011 (2011), `http://www.worldrobotics.org/`
80. Ikeda, A., Kurita, Y., Ueda, J., Matsumoto, Y., Ogasawara, T.: Grip force control for an elastic finger using vision-based incipient slip feedback. In: IEEE/RSJ International Conference on Intelligent Robots and Systems, Sendai, Japan, pp. 810–815 (2004)
81. INE: Comparación entre las poblaciones con y sin discapacidades (2008), `http://www.ine.es`
82. Jang, H., Moradi, H., Hong, S., Lee, S., Han, J.: Spatial reasoning for real-time robotic manipulation. In: IEEE/RSJ International Conference on Intelligent Robots and Systems, Beijing, China, pp. 2632–2637 (2006)
83. Johansson, R.S., Westling, G.: Roles of glabrous skin receptors and sensorimotor memory in automatic control of precision grip when lifting rougher or more slippery objects. Experimental Brain Research 56(3), 550–564 (1984)
84. Jorg, S., Langwald, J., Stelter, J., Hirzinger, G., Natale, C.: Flexible robot-assembly using a multi-sensory approach. In: IEEE International Conference on Robotics and Automation, vol. 4, pp. 3687–3694 (2000)
85. Kaneko, K., Kanehiro, F., Kajita, S., Hirukawa, H., Kawasaki, T., Hirata, M., Akachi, K., Isozumi, T.: Humanoid robot hrp-2. In: IEEE International Conference on Robotics and Automation, New Orleans, USA, vol. 2, pp. 1083–1090 (2004)
86. Katz, D., Brock, O.: Manipulating articulated objects with interactive perception. In: IEEE International Conference on Robotics and Automation, Pasadena, USA, pp. 272–277 (2008)
87. Kemp, C., Edsinger, A., Torres-Jara, E.: Challenges for robot manipulation in human environments. IEEE Robotics & Automation Magazine 14, 20–29 (2007)
88. Kemp, C.C., Anderson, C.D., Nguyen, H., Trevor, A.J., Xu, Z.: A point-and-click interface for the real world: laser designation of objects for mobile manipulation. In: ACM/IEEE International Conference on Human Robot Interaction, New York, NY, USA, pp. 241–248 (2008)
89. Khalil, W., Dombre, E.: Modeling identification and control of robots. Hermes Penton Science (2002)
90. Khatib, O.: A unified approach for motion and force control of robot manipulators: The operational space formulation. IEEE Journal of Robotics and Automation 3(1), 43–53 (1987)

91. Klingbeil, E., Saxena, A., Ng, A.Y.: Learning to open new doors. In: Robotics Science and Systems Workshop on Robot Manipulation (2008)
92. Kragic, D., Christensen, H.: Model based techniques for robotic servoing and grasping. In: IEEE/RSJ International Conference on Intelligent Robots and Systems, vol. 1, pp. 299–304 (2002)
93. Kröger, T., Finkemeyer, B., Thomas, U., Wahl, F.: Compliant motion programming: The task frame formalism revisited. In: Mechatronics & Robotics, Aachen, Germany (2004)
94. Lacey, G., Dawson-Howe, K.: The application of robotics to a mobility aid for the elderly blind. Robotics and Autonomous Systems 23(4), 245–252 (1998), http://www.ingentaconnect.com/content/els/09218890/1998/00000023/%00000004/art00011, doi:10.1016/S0921-8890(98)00011-6
95. Lasky, T., Hsia, T.: On force-tracking impedance control of robot manipulators. In: IEEE International Conference on Robotics and Automation, Sacramento, California, vol. 1, pp. 274–280 (1991), doi:10.1109/ROBOT.1991.131587
96. Latombe, J.C.: Robot Motion Planning. Kluwer Academic Publishers (1991)
97. LaValle, S., Yakey, J., Kavraki, L.: A probabilistic roadmap approach for systems with closed kinematic chains. In: IEEE International Conference on Robotics and Automation, vol. 3, pp. 1671–1676 (1999)
98. Lavalle, S.M.: Rapidly-exploring random trees: A new tool for path planning. Tech. Rep. TR 98-11. Computer Science Department, Iowa State University (1998)
99. Lee, M.H., Nicholls, H.R.: Tactile sensing for mechatronics–a state of the art survey. Mechatronics 9(1), 1–31 (1999)
100. Lee, S., Lee, S., Lee, J., Moon, D., Kim, E., Seo, J.: Robust recognition and pose estimation of 3d objects based on evidence fusion in a sequence of images. In: IEEE International Conference on Robotics and Automation, Rome, Italy, pp. 3773–3779 (2007)
101. Lepetit, V., Fua, P.: Monocular model-based 3d tracking of rigid objects. Foundations and Trends in Computer Graphics and Vision 1(1), 1–89 (2005)
102. Li, Z., Sastry, S.: Task oriented optimal grasping by multifingered robot hands. In: IEEE International Conference on Robotics and Automation, Raleigh, North Carolina, vol. 4, pp. 389–394 (1987)
103. Lowe, D.: Robust model-based motion tracking through the integration of search and estimation. International Journal on Computer Vision 8, 113–122 (1992)
104. Lynch, K.M., Murphey, T.D.: Control of Nonprehensile Manipulation. In: Lynch, K.M., Murphey, T.D. (eds.) Control Problems in Robotics, pp. 39–57. Springer, Heidelberg (2003)
105. Lyons, D.: A simple set of grasps for a dextrous hand. In: IEEE International Conference on Robotics and Automation, vol. 2, pp. 588–593 (1985)
106. Maass, J., Kohn, N., Hesselbach, J.: Open modular robot control architecture for assembly using the task frame formalism. International Journal of Advanced Robotic Systems 3(1), 1–10 (2006)
107. Maeda, Y., Arai, T.: A quantitative stability measure for graspless manipulation. In: IEEE International Conference on Robotics and Automation, vol. 3, pp. 2473–2478 (2002)
108. Maeda, Y., Nakamura, T., Arai, T.: Motion planning of robot fingertips for graspless manipulation. In: IEEE International Conference on Robotics and Automation, vol. 3, pp. 2951–2956 (2004)
109. Malis, E.: Survey of vision-based robot control. In: European Naval Ship Design, Captain Computer IV Forum. ENSIETA, Brest (2002)

110. Malis, E., Chaumette, F., Boudet, S.: 2 1/2 d visual servoing. IEEE Transactions on Robotics and Automation 15(2), 238–250 (1999)
111. Marchand, E., Chaumette, F.: Virtual visual servoing: a framework for real-time augmented reality. In: EUROGRAPHICS 2002, Saarebrücken, Germany, vol. 21(3), pp. 289–298 (2002)
112. Marrone, F., Raimondi, F., Strobel, M.: Compliant interaction of a domestic service robot with a human and the environment. In: 33rd Int. Symposium on Robotics, Stockholm, Sweden, pp. 7–11 (2002)
113. Martinet, P., Gallice, J.: Position based visual servoing using a nonlinear approach. In: IEEE/RSJ International Conference on Intelligent Robots and Systems, vol. 1, pp. 531–536 (1999)
114. Mason, M.: Compliance and force control for computer-controlled manipulators. IEEE Transactions on Systems, Man, and Cybernetics 11(6), 418–432 (1981)
115. Mason, M.: Progress in nonprehensile manipulation. International Journal of Robotics Research 18(11), 1129–1141 (1999)
116. Massimino, M.J., Sheridan, T.B.: Sensory substitution for force feedback in teleoperation. Presence: Teleoperators and Virtual Environments 2(4), 344–352 (1993)
117. Melchiorri, C.: Slip detection and control using tactile and force sensors. IEEE/SMC Transactions on Mechatronics 5(3), 235–243 (2000)
118. Mezouar, Y., Chaumette, F.: Optimal camera trajectory with image-based control. International Journal of Robotics Research 22(10), 781–804 (2003)
119. Miller, A.T., Allen, P.K.: Examples of 3d grasp quality computations. In: IEEE International Conference on Robotics and Automation, Detroit, Michigan, pp. 1240–1246 (1999)
120. Miller, A.T., Knoop, S., Christensen, H.I., Allen, P.K.: Automatic grasp planning using shape primitives. In: IEEE International Conference on Robotics and Automation, Taipei, Taiwan, pp. 1824–1829 (2003)
121. Morales, A., Asfour, T., Azad, P., Knoop, S., Dillmann, R.: Integrated grasp planning and visual object localization for a humanoid robot with five-fingered hands. In: IEEE/RSJ International Conference on Intelligent Robots and Systems, Beijing, China, pp. 5663–5668 (2006)
122. Morales, A., Sanz, P.J., del Pobil, A.P., Fagg, A.H.: Vision-based three-finger grasp synthesis constrained by hand geometry. Robotics and Autonomous Systems 54(6), 496–512 (2006)
123. Morel, G., Bidaud, P.: A reactive external force loop approach to control manipulators in the presence of environmental disturbances. In: IEEE International Conference on Robotics and Automation, Minneapolis, Minnesota, USA, vol. 2, pp. 1229–1234 (1996), doi:10.1109/ROBOT.1996.506875
124. Morel, G., Malis, E., Boudet, S.: Impedance based combination of visual and force control. In: IEEE International Conference on Robotics and Automation, Leuven, Belgium, vol. 2, pp. 1743–1748 (1998)
125. Morris, A., Donamukkala, R., Kapuria, A., Steinfeld, A., Matthews, J., Dunbar-Jacob, J., Thrun, S.: A robotic walker that provides guidance. In: IEEE International Conference on Robotics and Automation, vol. 1, pp. 25–30 (2003)
126. Mosemann, H., Wahl, F.: Automatic decomposition of planned assembly sequences into skill primitives. IEEE Transactions on Robotics and Automation 17(5), 709–718 (2001)
127. Murase, H., Nayar, S.: Learning and recognition of 3d objects from appearance. In: IEEE Workshop on Qualitative Vision, pp. 39–50 (1993), doi:10.1109/WQV.1993.262951

128. Napier, J.: The prehensile movements of the human hand. Journal of Bone and Joint Surgery 38-B(4), 902–913 (1956)
129. Napier, J.: Hands. Princeton University Press (1983)
130. Nelson, B., Morrow, J., Khosla, P.: Improved force control through visual servoing. In: American Control Conference, vol. 1, pp. 380–386 (1995)
131. Nelson, B., Khosla, P.K.: Force and vision resolvability for assimilating disparate sensory feedback. IEEE Transactions on Robotics and Automation 12(5), 714–731 (1996)
132. Nguyen, H., Kemp, C.C.: Bio-inspired assistive robotics: Service dogs as a model for human-robot interaction and mobile manipulation. In: 2nd IEEE RAS & EMBS International Conference on Biomedical Robotics and Biomechatronics, pp. 542–549 (2008), doi:10.1109/BIOROB.2008.4762910
133. Niao, S., Maruyama, Y.: Remote-operated robotic system for live-line maintenance work. In: IEEE International Conference on Transmission and Distribution Construction and Live-Line Maintenance, New York, USA, pp. 422–435 (1993)
134. Niemeyer, G., Slotine, J.J.: A simple strategy for opening an unknown door. In: IEEE International Conference on Robotics and Automation, Albuquerque, NM, USA, vol. 2, pp. 1448–1453 (1997)
135. OECD: OECD Factbook 2011–2012: Economic, Environmental and Social Statistics. OECD Publishing (2012)
136. Oestreicher, L., Severinson Eklundh, K.: User expectations on human-robot cooperation. In: The 15th IEEE International Symposium on Robot and Human Interactive Communication, pp. 91–96 (2006), doi:10.1109/ROMAN.2006.314400
137. Ohmura, Y., Kuniyoshi, Y.: Humanoid robot which can lift a 30kg box by whole body contact and tactile feedback. In: IEEE/RSJ International Conference on Intelligent Robots and Systems, San Diego, CA, USA, pp. 1136–1141 (2007)
138. Okamura, A.M., Smaby, N., Cutkosky, M.R.: An overview of dexterous manipulation. In: IEEE International Conference on Robotics and Automation, San Francisco, CA, USA, pp. 255–262 (2000)
139. Orebäck, A., Christensen, H.: Evaluation of architectures for mobile robotics. Autonomous Robots 14(1), 33–49 (2003)
140. Ott, C., Buml, B., Borst, C., Hirzinger, G.: Employing cartesian impedance control for the opening of a door: A case study in mobile manipulation. In: IEEE/RSJ International Conference on Intelligent Robots and Systems Workshop on Mobile Manipulators: Basic Techniques, New Trends & Applications, Edmonton, Canada (2005)
141. Perdereau, V., Drouin, M.: A new scheme for hybrid force-position control. Robotica 11, 453–464 (1993)
142. Pérez, J., Castellanos, J., Montiel, J.M., Neira, J., Tardós, J.: Continuous mobile robot localization: Vision vs. laser. In: IEEE International Conference on Robotics and Automation, Detroit, Michigan, USA, pp. 2917–2923 (1999)
143. Petersson, L., Austin, D., Kragic, D.: High-level control of a mobile manipulator for door opening. In: IEEE/RSJ International Conference on Intelligent Robots and Systems, Takamatsu, Kagawa, Japan, vol. 3, pp. 2333–2338 (2000)
144. Petersson, L., Egerstedt, M., Christensen, H.: A hybrid control architecture for mobile manipulation. In: IEEE/RSJ International Conference on Intelligent Robots and Systems, Kyongju, Korea (1999)
145. Petrovskaya, A., Khatib, O., Thrun, S., Ng, A.: Bayesian estimation for autonomous object manipulation based on tactile sensors. In: IEEE International Conference on Robotics and Automation, Orlando, FL, USA, pp. 707–714 (2006)
146. Petrovskaya, A., Ng, A.: Probabilistic mobile manipulation in dynamic environments with application to opening doors. In: Int. Joint Conf. on Artificial Intelligence, Hyderabad, India (2007)

147. del Pobil, A.P., Serna, M.A.: Spatial Representation and Motion Planning. LNCS, vol. 1014. Springer, Heidelberg (1995)
148. Pollard, N.S.: Parallel methods for synthesizing whole-hand grasps from generalized prototypes. Ph.D. thesis. Massachusetts Institute of Technology (1994)
149. Prats, M., Martínez, E., Sanz, P., del Pobil, A.P.: The uji librarian robot. Intelligent Service Robotics 1(4), 321–335 (2008)
150. Prats, M., Sanz, P.J., del Pobil, A.P., Martínez, E., Marín, R.: Towards multipurpose autonomous manipulation with the uji service robot. Robotica 25(2), 245–256 (2007)
151. Quigley, M., Berger, E., Ng, A.Y.: Stair: Hardware and software architecture. In: AAAI 2007 Robotics Workshop (2007)
152. Raibert, M., Craig, J.: Hybrid position/force control of manipulators. ASME Journal of Dynamic Systems, Measurement and Control 102(2), 126–133 (1981)
153. ReadyBot, `http://www.readybot.com/`
154. Rhee, C., Chung, W., Kim, M., Shim, Y., Lee, H.: Door opening control using the multi-fingered robotic hand for the indoor service robot. In: IEEE International Conference on Robotics and Automation, vol. 4, pp. 4011–4016 (2004)
155. Robonova, `http://www.robonova.com/`
156. Roy, N., Baltus, G., Fox, D., Gemperle, F., Goetz, J., Hirsch, T., Margaritis, D., Montemerlo, M., Pineau, J., Schulte, J., Thrun, S.: Towards personal service robots for the elderly. In: Workshop on Interactive Robots and Entertainment, WIRE 2000 (2000)
157. Sales, J., Garcia, X., Sanz, P., Marín, R., Prats, M., Falomir, Z.: Improving manipulation capabilities by means of radio frequency identification and computer vision. In: 11th International Conference on Climbing and Walking Robots and the Support Technologies for Mobile Machines (CLAWAR 2008), Coimbra, Portugal (2008)
158. Salisbury, J.: Active stiffness control of a manipulator in cartesian coordinates. In: IEEE International Conference on Decision and Control, Albuquerque, USA, pp. 95–100 (1980)
159. Salisbury, J.: Kinematics and force analysis of articulated hands. Ph.D. thesis. Standford University (1982)
160. Sanz, P., Prats, M., Ridao, P., Ribas, D., Oliver, G., Orti, A.: Recent progress in the RAUVI project. a reconfigurable autonomous underwater vehicle for intervention. In: 52th International Symphosium ELMAR 2010, pp. 471–474 (2010)
161. Sanz, P., Requena, A., Iñesta, J., del Pobil, A.P.: Grasping the not-so-obvious: vision-based object handling for industrial applications. IEEE Robotics & Automation Magazine 12(3), 44–52 (2005)
162. Sanz, P.J., Ridao, P., Oliver, G., Casalino, G., Insaurralde, C., Silvestre, C., Melchiorri, C., Turetta, A.: TRIDENT: Recent improvements about autonomous underwater intervention missions. In: International Federation of Automatic Control Workshop on Navigation, Guidance and Control of Underwater Vehicles, IFAC (2012)
163. Saxena, A., Driemeyer, J., Ng, A.Y.: Robotic grasping of novel objects using vision. International Journal of Robotics Research 27(2), 157–173 (2008)
164. Schlesinger, G.: Der mechanische Aufbau der kunstlichen Glieder in Ersatzglieder und Arbeitshilfen. Springer, Berlin (1919)
165. Schmid, A.J., Gorges, N., Göger, D., Wörn, H.: Opening a door with a humanoid robot using multi-sensory tactile feedback. In: IEEE International Conference on Robotics and Automation, Pasadena, CA, pp. 285–291 (2008)
166. Schraft, R., Schaeffer, C., May, T.: Care-o-bot: the concept of a system for assisting elderly or disabled persons in home environments. In: 24th Annual Conference of the IEEE Industrial Electronics Society, vol. 4, pp. 2476–2481 (1998), doi:10.1109/IECON.1998.724115

167. Schulenburg, J.: Gocr, `http://www-e.uni-magdeburg.de/jschulen/ocr/`
168. Se, S., Lowe, D., Little, J.: Vision-based mobile robot localization and mapping using scale-invariant features. In: IEEE International Conference on Robotics and Automation, Seoul, Korea, pp. 2051–2058 (2001)
169. Sheridan, T.: Telerobotics, Automation and Human Supervisory Control. MIT Press (1992)
170. Siciliano, B.: Kinematic control of redundant robot manipulators: A tutorial. Journal of Intelligent and Robotic Systems 3(3), 201–212 (1990)
171. Siciliano, B., Villani, L.: Robot force control. Kluwer Academic Publ., Boston (2000)
172. Siegwart, R.: Robox at Expo.02: A Large Scale Installation of Personal Robots. Robotics and Autonomous Systems 42(3-4), 203–222 (2003), doi:NA
173. Son, J.S., Howe, R., Wang, J., Hager, G.: Preliminary results on grasping with vision and touch. In: IEEE/RSJ International Conference on Intelligent Robots and Systems, Osaka, Japan, vol. 3, pp. 1068–1075 (1996)
174. Stanger, C., Anglin, C., Harwin, W., Romilly, D.: Devices for assisting manipulation: a summary of user task priorities. IEEE Transactions on Rehabilitation Engineering 2(4), 256–265 (1994), doi:10.1109/86.340872
175. Stansfield, S.: Robotic grasping of unknown objects: A knowledge-based approach. International Journal of Robotics Research 10(4), 314–326 (1991)
176. Stemmer, A., Schreiber, G., Arbter, K., Albu-Schäffer, A.: Robust assembly of complex shaped planar parts using vision and force. In: IEEE International Conference on Multisensor Fusion and Integration, Heidelberg, Germany, pp. 493–500 (2006)
177. Stilman, M.: Task constrained motion planning in robot joint space. In: IEEE/RSJ International Conference on Intelligent Robots and Systems, pp. 3074–3081 (2007)
178. Suthakorn, J., Lee, S., Zhou, Y., Thomas, R., Choudhury, S.: A robotic library system for an off-site shelving facility. In: IEEE International Conference on Robotics and Automation, Washington, DC, USA, pp. 3589–3594 (2002)
179. Taylor, C.L., Schwarz, R.J.: The anatomy and mechanics of the human hand. Artificial Limbs 2(2), 22–35 (1955)
180. Taylor, G., Kleeman, L.: Flexible self-calibrated visual servoing for a humanoid robot. In: Proc. of the Australian Conference on Robotics and Automation, Sydney, Australia, pp. 79–84 (2001)
181. Tegin, J., Wikander, J.: Tactile sensing in intelligent robotic manipulation - a review. Industrial Robot: An International Journal 32(1), 64–70 (2005)
182. Thomas, U., Finkemeyer, B., Kröger, T., Wahl, F.: Error-tolerant execution of complex robot tasks based on skill primitives. In: IEEE International Conference on Robotics and Automation, Taipei, Taiwan, vol. 3, pp. 3069–3075 (2003)
183. Thuilot, B., Martinet, P., Cordesses, L., Gallice, J.: Position based visual servoing: keeping the object in the field of vision. In: IEEE International Conference on Robotics and Automation, Washington DC, USA, pp. 1624–1629 (2002)
184. Tomizawa, T., Ohya, A., Yuta, S.: Remote book browsing system using a mobile manipulator. In: IEEE International Conference on Robotics and Automation, Taipei, Taiwan, pp. 256–261 (2003)
185. Twendy-one, `http://twendyone.com/`
186. Vasudevan, S., Gachter, S., Nguyen, V., Siegwart, R.: Cognitive maps for mobile robots – an object based approach. Robotics and Autonomous Systems 55, 359–371 (2007)
187. Vestka, L.E.K.P., Claude Latombe, J., Overmars, M.H.: Probabilistic roadmaps for path planning in high-dimensional configuration spaces. IEEE Transactions on Robotics and Automation 12, 566–580 (1996)

188. Weiss, K., Wörn, H.: Tactile sensor system for an anthropomorphic robot hand. In: IEEE International Conference on Manipulation and Grasping, Genoa, Italy (2004)
189. Whitney, D.: Force feedback control of manipulators fine motions. ASME Journal of Dynamic Systems, Measurement, and Control, 91–97 (1977)
190. Wren, D., Fisher, R.: Dextrous hand grasping strategies using preshapes and digit trajectories. In: IEEE International Conference on Systems, Man and Cybernetics, Vancouver, BC, Canada, vol. 1, pp. 910–915 (1995)
191. Wyrobek, K., Berger, E., Van der Loos, H., Salisbury, J.: Towards a personal robotics development platform: Rationale and design of an intrinsically safe personal robot. In: IEEE International Conference on Robotics and Automation, Pasadena, California, pp. 2165–2170 (2008)
192. Yao, Z., Gupta, K.: Path planning with general end-effector constraints: using task space to guide configuration space search. In: IEEE/RSJ International Conference on Intelligent Robots and Systems, pp. 1875–1880 (2005)
193. Zhu, X., Wang, J.: Synthesis of force-closure grasps on 3-d objects based on the q distance. IEEE Transactions on Robotics and Automation 19(4), 669–679 (2003)
194. Zumel, N., Erdmann, M.: Nonprehensile manipulation for orienting parts in the plane. In: IEEE International Conference on Robotics and Automation, vol. 3, pp. 2433–2439 (1997)

MIX
Papier aus verantwortungsvollen Quellen
Paper from responsible sources
FSC® C105338

If you have any concerns about our products,
you can contact us on
ProductSafety@springernature.com

In case Publisher is established outside the EU,
the EU authorized representative is:
Springer Nature Customer Service Center GmbH
Europaplatz 3, 69115 Heidelberg, Germany

Printed by Libri Plureos GmbH
in Hamburg, Germany